Designing High-Performance Stiffened Structures

Based on papers presented at the one-day seminar *Designing High Performance Stiffened Structures*, held at the IMechE Headquarters, London, UK, on 29 June 1999.

IMechE
Seminar Publication

I MECH E

Designing High-Performance Stiffened Structures

Organized by
The Aerospace Industries Division of
The Institution of Mechanical Engineers

Co-sponsored by
The Automobile Division and the Materials and Mechanics of Solids Group of
The Institution of Mechanical Engineers

IMechE Seminar Publication 2000–11

**Professional
Engineering
Publishing**

Published by Professional Engineering Publishing Limited for The Institution of
Mechanical Engineers, Bury St Edmunds and London, UK.

First Published 2000

© 2000 The Institution of Mechanical Engineers, unless otherwise stated.

ISSN 1357–9193
ISBN 1 86058 308 3

A CIP catalogue record for this book is available from the British Library.

Printed by The Cromwell Press, Trowbridge, Wiltshire, UK.

Related Titles of Interest

Title	Editor/Author	ISBN
Finite Element Analysis of Elastomers	D Boast and V A Coveney	1 86058 171 4
Creep of Materials and Structures	T H Hyde	0 85298 900 8
IMechE Engineers' Data Book, E2	C Matthews	1 86058 248 6
Design of Composite Structures Against Fatigue: Applications to Wind Turbine Blades	R M Mayer	0 85298 957 1
Evaluating Materials Properties by Dynamic Testing	E van Walle	1 86058 004 1
Impact and Dynamic Fracture of Polymers and Composites	J G Williams and A Pavan	0 85298 946 6
Designing Cost-Effective Composites	IMechE Conference	1 86058 148 X
Joining and Repair of Plastics and Composites	IMechE Conference	1 86058 198 6
Composites for the Offshore Oil and Gas Industry	IMechE Seminar	1 86058 229 X

For the full range of titles published by Professional Engineering Publishing contact:

Sales Department
Professional Engineering Publishing Limited
Northgate Avenue
Bury St Edmunds
Suffolk
IP32 6BW
UK

Tel: +44 (0)1284 724384
Fax: +44 (0)1284 718692

Contents

RS-3 polycyanate ester resin – the satellite's community choice for high-performance composites

S ROBITAILLE, G PATZ, and **M SABA**
YLA Inc Advanced Composite Materials, Benicia, USA

INTRODUCTION

Since the mid 1980's cyanate ester (CE) or polycyanurate resin chemistry have been characterized and qualified into many advanced composite structures and applications. The greatest advantages of this class of thermosetting resin have been found in the support of electronic and space structures. The unique attributes of the cyanate ester resin chemistry and resulting properties has provided advantages over other characterized thermoset resins. While by far the largest usage of cyanate ester chemistry is in the circuit board industry, with over 60 % of the entire CE market, this paper will focus on their use in space structures for stable structures and low dielectric applications such as radomes and antennas.

Generally the most attractive attributes of CE chemistry evolve from the cured chemical structure. While there are differences in performance depending on the backbone structure and formulation of CE's, they all contain a notably low concentration of dipoles and hydroxyl groups in the cured structure. The result is lower moisture absorption, higher diffusivity and low dielectric constant and loss when compared with epoxy and BMI systems. These attributes are particularly attractive for stable structures and low dielectric applications.

The CE cure process involves polycyclotrimerization of three OCN groups to form a triazine ring structure. This structure offers good modulus retention and glass transition temperatures in the 220°C range with some exceptions up to 300°C as is the case of the phenol triazine, or PT resins.

Representative Cure Chemistry of Dicyanates

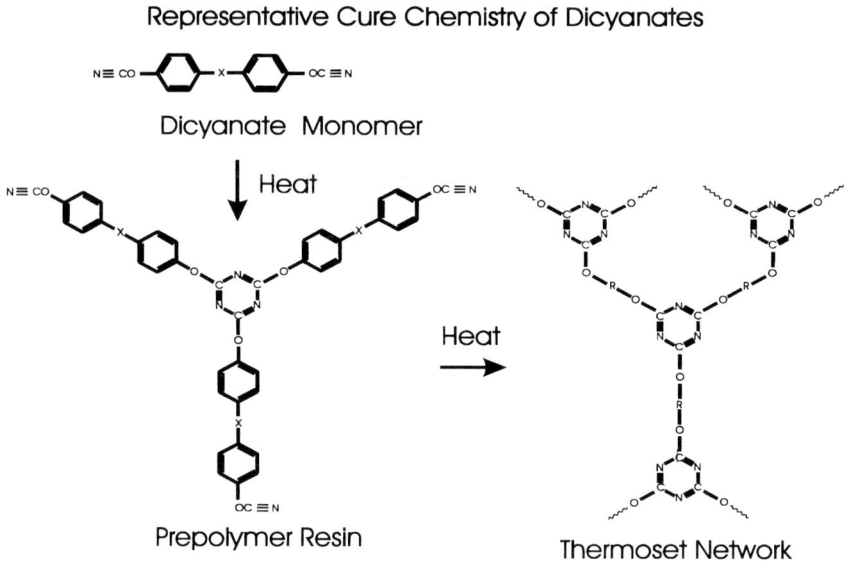

Figure 1. Cure reaction of polycyanate resins

The reaction does not generate volatile components during cure and they are processable using hot melt impregnation methods.

The resin modulus and toughness characteristics depend in part on the backbone structure and crosslink density of the polymer. CE is typically toughened by conventional mechanisms used with epoxy and BMI resins with the expected balance of properties. The ability to modify and toughen CE based resins makes them appropriate for adhesives and toughened composite applications.

MATERIAL PROCESSING AND PROPERTIES

CE have similar composite processing characteristic to epoxy resins. They are available as hot melt prepregs, RTM resins, adhesive systems and syntactic core materials. They can be consolidated using autoclave, vacuum bag, press, pultrusion, and RTM methods. In order to achieve adequate conversions of the OCN reactive groups and useable room temperature out times, most cure temperatures range from 120°C to 180°C. It is recommended that for structural service temperatures of 180°C, low dielectric properties or for improved hydrolytic and chemical stability that the CE resins be postcured to 235°C - 315°C.

The rheology curve Figure. 2., shows a comparison between a non-flow controlled toughened cyanate ester resin (RS-3) and a 180°C curing TGDMA/ DDS epoxy used for aerospace structural applications.

S695/001/99

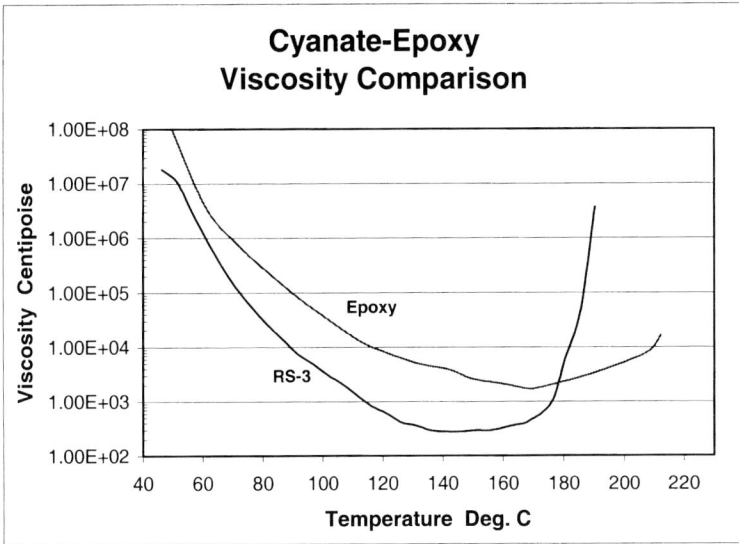

Figure. 2. Rheology curve of RS-3 Cyanate ester and TGMDA/DDS epoxy.

The minimum viscosity of the CE resin occurs at lower temperatures than the epoxies tested but are easily adjusted for flow using thixotropes and thermoplastic modifications. This minimum viscosity and large processing window allows the CE resin to be processed by a variety of methods, from vacuum bag curing to RTM and resin film infusion.

CE REACTION WITH MOISTURE:

As in most high performance composites processing, exposure to contamination during the cure must be controlled. In the case of CE by far the most predominant contamination problem occurs when the resins are exposed to moisture during the cure process. Carbamate contamination of cyanate ester systems due to the reaction with moisture in structural composites has been noted since the early 1990's, when the first structural parts were fabricated for space structures using low CTE composite tooling. Shimp, et al, has characterized the reaction. The general reaction for the formation of carbamate is shown below: [1]

$$R-O-C{\equiv}N + H_2O \Rightarrow R-O-\underset{\underset{N\ H}{\|}}{C}-H \Rightarrow R-O-\underset{\underset{O}{\|}}{C}-NH_2$$

carbamate

$$R-O-\underset{\underset{O}{\|}}{C}-NH_2 \Rightarrow R-NH_2 + CO_{2\ gas}$$

$$R-NH_2 + R-O-C{\equiv}N \Rightarrow R-NH-\underset{\underset{NH}{\|}}{C}-O-R$$

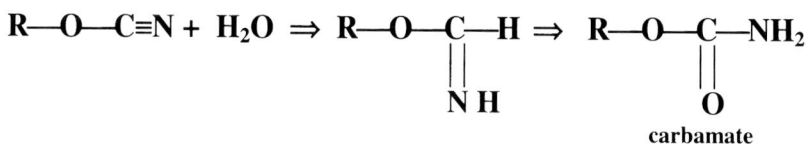

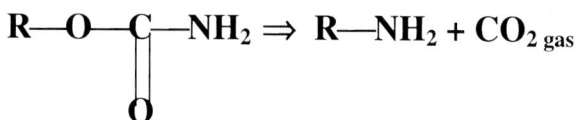

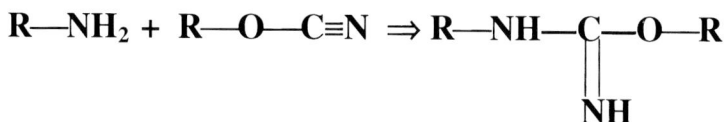

**Figure 3. Formation of carbamate and isourea due to reaction
with moisture in cyanate ester resin.**

The reaction of water and cyanate esters proceeds very slowly at room temperature and is generally not an issue as long as good material storage practices are used. The formation of the carbamate can occur when moisture is able to diffuse into the laminate or adhesive and react with the polymer system during the cure process. This problem is most noticeable on structures that are cured against undried composite tools or moisture laden bagging materials, fibers and foam cores. The formation of the carbamate group proceeds by hydrolysis of the OCN functional group to form a carbamate. With additional heating to temperatures above 150 °C, the carbamate group decomposes and forms an amine and carbon dioxide. The carbon dioxide gas can cause blistering and delamination as noted in early post cured Kevlar laminates by Shimp and Ising and was dependent on the catalyst used in the formulation. During this decomposition to carbon dioxide if there are still unreacted OCN groups available, the amine will react and form a more stable linear polymer structure, or isourea. [1, 2]

Moisture contamination problems do not always result in blistering or delamination. For composites cured against un-dried composite tooling the problem generally is restricted to the tool surface of the laminate. In severe cases it can be detected as a rough, friable surface on laminates. If the formation and/ or decomposition of the carbamate structure has occurred the surface of the laminate should be soluble in a ketone solvent such as acetone or methyl ethyl ketone. We have performed many surface extractions and FTIR analysis on suspect laminates and compared these with controls. Depending on the cure conditions, FTIR examination of the extract can detect the carbonyl peak at 1750 cm-1. However this can be misleading, if the carbamate has decomposed fully due to elevated cure temperatures and is no longer present, you may only detect a much stronger than normal NH stretch peak around 3500cm-1 or even the isourea peak at 2360 cm-1. One characteristic change of the cured cyanate if carbamates have been formed is a decrease in the glass transition temperature of the resin. Using the decrease in Tg to determine if there has been reaction with water may only be useful in the composite if the contamination is gross and effects the entire laminate, not just a surface contamination.

Another useful tool to determine the potential presence of carbamates and their decomposition on suspect parts is detailed Electron Spectroscopy for Chemical Analysis or ESCA which is used for determining elemental and chemical bonding information for elements which have atomic numbers greater than helium. This method can be used to determine the existence of carbamate groups or the amine and subsequent isourea structure formed from carbamate decomposition and other potential surface contaminates like silicone and fluorine.

The selection of the metal catalyst used for the CE systems is an important factor in limiting the generation of the carbamate group and the sensitivity of the resin system to hydrolysis. Shimp determined that the metal chelates of copper and colbalt resulted in the best resistance to hydrolysis and the formation of carbamates. Hydrolytic and chemical resistance is also dependent on % conversion of the resin and the actual CE resin chemistry used in the formulation. [3]

In order to eliminate and alleviate the risk of forming carbamates during cure, some easy precautions and techniques can be used. These include pre-drying and dry storage of composite tools, drying and outgassing foam cores, drying fabrics and modifying the cure cycle to remove moisture from materials, such as Kevlar, before the cure reaction begins. Even changes in the vacuum bag lay-up, allowing a good path for trapped moisture to be removed from the vacuum bag and keeping the part under an active vacuum thorough out the cure have been found to be useful in minimizing the problem.

CE PROPERTIES

It is difficult to realistically compare CE and epoxy resins due to the availability of a wide range of polymer properties for both systems. Though for comparison sake we have selected a standard aerospace system based on TGMDA / DDS and a CE qualified to the majority of aerospace applications using CE resins. It is based on the dicyclopentadiene backbone CE from Ciba Giegy. Comparing the neat resin properties of CE and epoxy, the characteristic properties of CE give lower moisture absorption, CTE and dielectric properties than epoxy system. CE's also fill the gap in thermal properties between epoxies and BMI with good retention of wet Tg.

Table 1. Compares Typical neat resin properties for 180°C curing TGMDA / DDS epoxy and 180°C Cyanate Esters

Neat Resin Properties	Polycyanates	Epoxies
Density g/cc	1.19 - 1.24	1.25 - 1.30
Tg (C)	200 - 315	180 - 220
Wet Tg (C)	180 - 305	130 - 180
Flexural Strength ksi	12.5	10. - 15
Flexural Mod, msi	0.4	0.6
Tensile Str, ksi	11.6	12
Tensile Mod, msi	0.45	0.6
Tensile Strain %	2.0 - 4.9	0.7 - 3
CTE in/in F (-212 - +300 F)	24	27
TML%	0.1 - 0.3	0.3 - 1.5
CVCM%	0 - 0.01	0.01 - 0.8
Moisture Absorption %	0.6 - 2.5	4.0 - 7
Dielectric Properties		
Dk	2.67- 2.78	3.5 - 4.0
Df	0.003 - 0.008	0.04

Table 2

RS-3 / G-40-800 Uni-tape Properties				
350 F/ 2 hrs/ 450 2 hr PC				
G-40 - 800 / RS-3	RTD	RTW	177C dry	177C wet
[0] Flex Str MPa (ksi)	1407 (204)	1297 (188)	979 (142)	869 (126)
[0] Flex mod.GPa (msi)	144 (21)	142 (20.6)	141 (20.4)	139 (20.1)
[0] Flex Strain%	0.984	0.927	0.693	0.63
[0] Tensile str., MPa(ksi)	322 (467)	318 (461)	286 (415)	287 (416)
[0] Tenisle mod. GPa (msi)	170 (24.7)	164 (23.8)	161 (23.4)	162 (23.4)
[0] Tensile Strain %	1.78	1.86	1.76	1.82
ILSS	86.2 (12.5)	82.8 (12.0)	59.3 (8.6)	53.8 (7.8)
[0] Comp Str. Mpa (ksi)	1413 (205)	1517 (220)	1120 (162)	869 (126)
[0] Comp Mod. GPa (msi)	151 (22.0)	161 (23.3)	146 (21.2)	141 (21.9)
G_{IIC} KJ/m^2	1.2			
EDS Str. Mpa (ksi)	287 (41.6)			
OHT Mpa (ksi)	601 (87.2)			
OHC Mpa (ksi)	310 (44.9)			
In-Plane Shear Str. GPa (ksi)	62.1 (9.0)		35.2 (5.1)	
CAI BMS 7260 1500 in-lb/in	215 (31.2)			
Wet Cond 76.7 C 90% RH, eq moisture 0.273%				

S695/001/99

Table 2. Show mechanical properties of RS-3, a toughened cyanate ester resin on G-40-800 fiber. Note the excellent balance and retention of properties at elevated temperature and wet conditions. RS-3 uses a proprietary sub micron core shell toughener that toughens the polymer by inducing shear yielding at relatively low concentrations. This toughening method results in almost no loss of Tg when compared to the unmodified polymer and is independent of phase separation during cure. [4]

CE SYSTEM STABILITY

There are several areas that CE chemistry offers better structural stability when compared with epoxy systems. These characteristics have made them successful systems for space applications where structural stability is more important that in other aerospace applications. CE's generally have lower cure shrinkage, CTE, CME, and microcracking than epoxy systems. Their ability to perform well mechanically on ultra high modulus pitch fibers with a more negative CTE allows the design of structures with a near zero CTE over a wide thermal range. In addition, toughened formulations show excellent resistance to cryogenic temperatures and thermal shock. [5]

The coefficient of moisture expansion, CME, of composite structures is an important design variable for satellite structures. The factors influencing the behavior are generally accepted to be the amount of moisture absorbed by the resin and in induced strain resulting from the absorption.

Moisture Uptake Behavior of Cyanate Ester and Epoxy Neat Resin
Exposure 100% R.H. 75 deg. F, > 1000 days

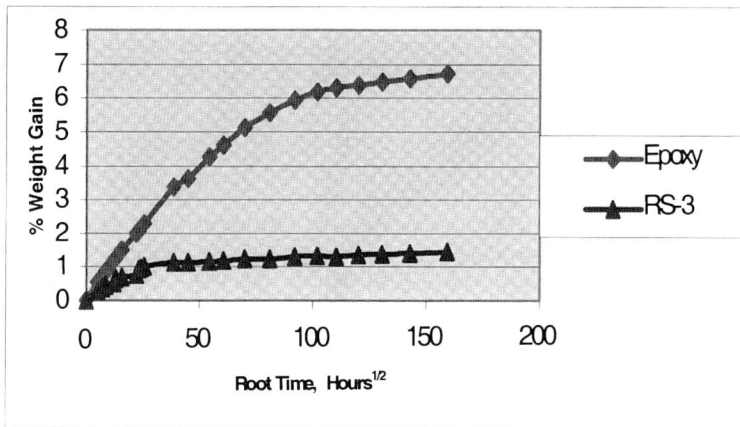

Figure. 4. Moisture absorption of CE compared with 180°C curing epoxy system. [6]

The moisture absorption of behavior of RS-3 is consistent with other CE resins. Equilibrium moisture content is reached quickly and stabilizes at relatively low levels.   There is speculation that the physical mechanism is not wholly based on Fickian behavior but maybe associated with a volumetric mechanism and the availability of free volume in the cured resin [7,8]. The measured moisture diffusivity of CE is generally higher than epoxy systems. Using the rational that the resin swelling is caused by water molecules clustering around strong polar groups, then you would expect that only about 20 % of the total moisture absorbed by CE resin to be associated with dipoles [7]. The high coefficient of diffusivity would tend to agree with the lack of strong polar bonding with water and favor a free volume mechanism resulting in less moisture induced swelling. The induced strain is slightly lower than epoxy systems tested approximately 600 ppm for the neat resin saturated to 0.44%. This results in an overall lower CME in the composite and very low outgassing values. [9]

Comparison of Wet Tg
Neat Resin

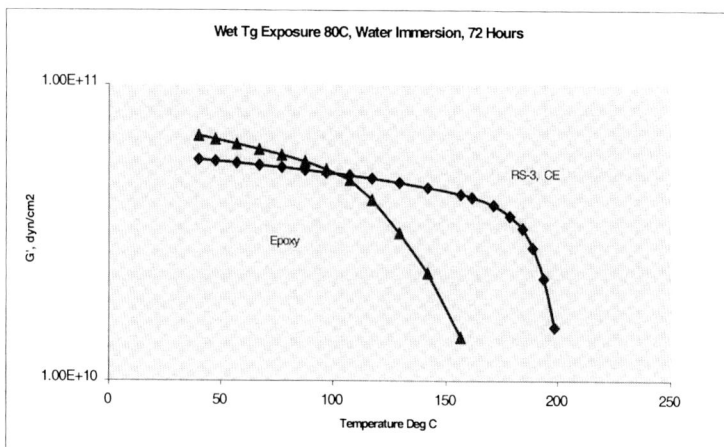

Wet Tg Exposure 80C, Water Immersion, 72 Hours

Figure. 5. Plot of the G' shear modulus of wet conditioned neat resin castings, 0.060"
thick.

The low moisture absorption and high diffusivity of CE is also consistent with the retention of Tg under wet conditions. In Figure 5, We measured the shear modulus for both resins after being exposed to the same moisture conditioning. While their dry Tg are similar both around 200°C the wet Tg retention of the systems vary greatly.

MICROCRACKING

For stable structures applications the ability to predict and limit composite microcracking is an important criteria for designing structure stability. Thermal cycling under vacuum to establish the dimensional and mechanical performance change a structure will exhibit is currently performed for many space structures prior to launch. The ability of a composite system to reduce the effect of microcracking will result in limited dimensional and mechanical performance change over the predicted lifetime of the structure. CE's have shown that over many cycles, to resist microcracking at cryogenic temperatures both as adhesives and composites. See Table 3. But under extreme thermal cycling ± 120° C > 1000 cycles the microcracks that did form had little effect on the mechanical and CTE performance of the structure. Studies conducted by SS Loral, Hughes Space and Communications, Lockheed Martin and Nippon Graphite have all shown RS-3 gave improved resistance to the effects of microcracking over other thermosetting systems. [5, 9]

Table 3. Observed microcracking by thermal cycling.

CE low temperature data courtsey of Lockheed Martin.

Material ID	Description	Size	Conditioning	Cycles	Microcrack Analysis
M133 laminate	T-800/RS-3	5"X.5"X0.090"	-423 to RTD to +250F	50	No failure
M88 neat resin casting	RS-4A	3"X.5"X0.080"	-423 to RTD to +250F	50	No failure
M123 neat resin casting	RS-4A paste	3"X0.5"X0.080"	-423 to RTD to +250F	50	No failure
Composite lap shear	RS-4A		-423 to RTD to +250F	50	No failure
Composite lap shear	RS-4A paste		-423 to RTD to +250F	50	No failure

Failure analysis: visual, microscopic, 5X, 50X, 100X, 400X

Thermal cycling of both CE adhesive, neat resin and composites RS-3 / T-800 (0,90) showed no visual microcracking after 50 cycles, Table 3. Cycle conditioning was performed by placing room temperature specimens in a -423°F environment for 10 minutes, removing the specimens until they reached room temperature, then placing them in a 250 °F oven for 10 minutes. After 50 cycles the samples were micrographed and compared to uncycled specimens to determine whether they showed any microcracks, none were detected using microscopic evaluation. (RS-4A is a 180°C CE adhesive system qualified for space applications). [5]

DIELECTRIC PROPERTIES

The exposure of an insulating material, as most unmodified thermosetting polymers are, to an electromagnetic field will result in absorption and storage of energy by the material and dissipation of a portion of that energy as heat. This absorption is characteristic of the dielectric loss for that material. For microwave windows such as radomes and high power antenna structures a high loss material will result in a loss of signal, signal distortion, a decrease in power and an increase in part temperature. For advanced high power radar systems the generation of heat can be significant and can result in part failure. Most of the data we are familiar with for dielectric properties required for antenna and radome systems is performed in the X band frequency range however, a study performed by Speak on CE quartz laminates showed excellent stability for both dielectric constant and loss values from X band through W band. [10]

As we have previously discussed for cyanate ester resins, the symmetrical structure of the cured polymer results in a high degree of charge balance, short dipole moments and low polarity. Unlike epoxy resins, the cure process of CE resins does not produce hydroxyl groups, This provides few sites for strong hydrogen bonding, reducing the polarity of the resin and decreasing the dielectric values. Differences in backbone structures and percent conversion can also effect the dielectric properties but in general the cured chemical structure of CE's support the observed low dielectric constant and loss data. [11].

Table 4. Comparison data of dielectric properties.

Composite : 12- 18 GHz	Dielectric Constant	Dielectric Loss
CE/ Quartz Laminate	3.28	0.005
CE/ Spectra	2.7	0.005
CE syntactic, 0.64 g/cc	1.7	0.004
BMI/ Quartz Laminate	3.55	0.016
CE Resin (RS-3)	2.76	0.005
Epoxy/ Quartz Laminate	3.75	0.018

The low moisture absorption of CE resins also results in improved performance in electrical applications. Lower overall moisture absorption of CE reduces the effect of exposure on the dielectric loss properties.

CONCLUSION

For satellite, radome and antenna structures CE resin chemistry has provided greater structural stability, weight savings and improved performance over many metal and epoxy composite materials. CE chemistry offers excellent advantages over epoxy resin systems but as with most materials there are trade–offs. The costs of CE resins are high in comparison to epoxy and BMI resins. Resin costs range between $ 60 - $120 UDS per lb. Their relative newness in the composite industry, compared to epoxy and BMI resins mean that a complete and broad database for applications outside space, circuit boards, and some advance radome applications is limited.

Although higher in material cost the improved performance of CE resins continues to results in increased usage in space applications, stable structures and advanced radome systems. Estimates of resin usage for these specific applications are approximately 25,000 lbs. per year. Future usage is expected to grow 20 % per year. [12]

REFERENCES

1. J.P.Pascault, J. Galy and F. Mechin, (1994) *Chemistry and Technology of Cyanate Ester Resins*, Hamerton, 117

2. Shimp, D.A. and Ising, S. J. (1991) *Am. Cem. Soc. Polyn Sci. Eng. Preprints*, 66, 504

3. Shimp, D.A. and Southcott, M. (1993) *Proceedings of International SAMPE Symposium Vol. 38,* 370

4. Yang, P.C., Woo, E.P., Laman, S.A., Jakubowski, J.J. (1991) *Proceedings of International SAMPE Symposium., Vol. 36.,* 437

5. Stift, M., Sidwell, D. (1999) *Out of Autoclave, 250°F Maximum Cure Composite Resin System Evaluation For Cryogenic Applications,* Report, Lockheed Martin

6. *Space Applications Granoc Pitched Based Fibers*, (1995) Data Package

7.   D.A. Shimp and B. Chin. (1994) *Chemistry and Technology of Cyanate Ester Resins*, Hamerton , 243

8.   V. Bellenger, J. Verdu, E. Morel (1989) Journal of Materials Science, 24,   63- 68

9.   Materials Directorate, Wright Laboratory, WPAFB, (1997), *High Modulus   Prepreg Development*

10.   S.C. Speak, H. Sitt, R.H. Fuse, (1991) *Proceedings of International SAMPE Symposium.*, Vol. 36.,   336

11.   D.A. Shimp and B. Chin. (1994) *Chemistry and Technology of Cyanate Ester Resins*, Hamerton, 231

12.   Verbal Conference, Ciba Giegy

S695/002/99

Analysis of the buckling behaviour of stiffened prismatic structures

D J DAWE
School of Civil Engineering, The University of Birmingham, UK

1   INTRODUCTION

The finite strip method (FSM) is a powerful and effective procedure for the numerical analysis of the behaviour of both single rectangular plates (flat and curved) and complicated prismatic plate and shell structures. Fig. 1 shows an example of a flat plate structure modelled by a number of finite strips which run the whole length of the structure. This feature, of the strips running along the complete structure, is a distinguishing feature of the FSM as compared to the general finite element method (FEM) but, in essence, the FSM is a specialised form of the FEM. Clearly the FSM is less versatile than is the FEM but in situations where it can be applied, as in analysing prismatic structures, it can be more powerful and economical, sometimes to a very marked extent.

In modelling a structure the finite strips are rigidly joined at their external longitudinal reference lines at which the requisite compatibility conditions are satisfied. With regard to Fig. 1 it should be noted, though, that finite strips usually have reference lines (i.e. lines with which degrees of freedom are associated) in their interior, in addition to at their exterior longitudinal edges, and that in some types of analysis the length of the strips which is considered for analysis purposes is less than the full length of the structure (e.g. it may be one half-wavelength of a buckling mode).

In the analysis of composite laminated structures it is necessary to take account of the very general material properties that may occur, such as anisotropy and coupling between in-surface (membrane) and out-of-surface (bending) actions. Furthermore, in considering the out-of-surface behaviour of component plates it is often necessary to take account of through-thickness shearing effects. The classical theory ignores these effects, of course, whilst the first-order shear deformation theory is the simplest theory that includes them, to provide a more realistic (but more complicated) model for out-of-surface behaviour. Both kinds of theory are of interest here. It is noted that higher-order shear deformation theories do exist for compact laminates but their use introduces much extra complexity without a corresponding gain. However, for sandwich construction a higher-order theory is appropriate.

This paper is concerned with describing the use of the finite strip method in predicting the behaviour of prismatic structures. A description is given first of the use of the method in the fundamental problem of predicting the buckling stresses of single-span, composite laminated plate and shell structures. Thereafter, other categories of problem are considered briefly. These comprise the buckling of multi-span structures with possible longitudinal thickness change, thermal buckling, the buckling of sandwich structures, aspects of post-buckling behaviour, and vibration and dynamic response.

The work described here has been conducted at The University of Birmingham over a considerable period of time. The various aspects of the work can only be summarised here, without any technical detail, but full details are available in the appended selected references. It is appropriate, and a pleasure, to record here the names of colleagues who have contributed to this work over the years: these are Onsy Roufaeil, Ian Morris, Trevor Craig, Zaven Azizian, Vahigh Peshkam, Suk Wai Fong, Siu-Shu Lam, Sabarudin Mohd, Shemin Wang, Jiye Chen, Weixing Yuan, Dongyao Tan and Yong Sheng Ge. It is also pertinent to note, with thanks, that financial support for the work has been provided by the EPSRC, by DERA Farnborough and by DERA Rosyth, and that other support has come from BAe Warton and Lloyd's Register of Shipping.

2   BUCKLING OF LAMINATED PLATE AND SHELL STRUCTURES

2.1   The finite strip

A particular type of individual curved finite strip is shown in Fig. 2(a) : the strip has four longitudinal reference lines and this corresponds to cubic polynomial interpolation of displacement quantities around the strip. The local axes x, y and z are surface ones, i.e. are axial, circumferential and normal ones.   The corresponding perturbation translational displacements at the middle surface are u, v and w. In shear deformation shell theory (SDST) analysis the rotations ψ_x and ψ_y of the middle-surface normal along the x and y directions, respectively, are further independent quantities, while in thin shell theory (TST) analysis these rotations are not present as independent quantities. It is noted that the description that follows relates primarily to SDST analysis but a corresponding TST analysis is also always available. It is also noted that when referring to flat plate theory, SDPT and CPT denote shear deformation plate theory and classical plate theory. The various detailed aspects of the work described in this Section are given in Refs [1-9].

The finite strip may be subjected to an applied stress system, comprising σ_x°, σ_y° and τ_{xy}° as shown in Fig. 2(b), leading to buckling. Each of the applied stresses is taken here to have uniform distribution throughout the strip, as illustrated, but it is possible to accommodate non-uniform distributions.

2.2   Energy and strip matrices

The required properties of a finite strip for the buckling problem are encapsulated in an elastic stiffness matrix, **k**, and a geometric stiffness matrix, $\mathbf{k_g}$. The stiffness matrix is based on the expression for strip strain energy and this is further based on the use of the enhanced Koiter-

Sanders SDST equations. The geometric stiffness matrix is based on expressions for the potential energy of the applied stress system. The energy expressions are not presented here but it is noted that they are comprehensive. Thus the laminate constitutive equations used in the strain energy expression are complete (including all anisotropy and coupling coefficients) and the potential energy of the applied stress system includes all de-stabilising terms associated with both in-surface and out-of-surface instability.

2.3 Strip displacement fields

The evaluation of the strip matrices requires the use of an appropriate displacement field substituted into the relevant energy expressions and the selection of the field is of great importance, of course. We need to distinguish between the modelling of actual, finite-length structures and of "long" structures, and between the displacement fields of two variants of the FSM, namely the semi-analytical finite strip method (S-a FSM) and the B-spline finite strip method (B-s FSM, or simply spline FSM).

2.3.1 Multi-term analysis of finite-length structures

Here it is generally necessary to employ a strip displacement field of the multi-term type, i.e. a field in which each of the displacement reference quantities (u, v, w, ψ_x and ψ_y) is represented as a series of products of longitudinal functions and crosswise, or circumferential, functions. The displacement fields for the multi-term S-a FSM and the multi-term B-s FSM have the same kind of crosswise representation in both approaches, which is based on the use of standard polynomial shape functions of the type routinely used in finite element analysis. The degree of the shape functions can be varied to create different types of strip model, corresponding to linear or quadratic or cubic or higher-degree polynomial representation. The finite strip shown in Fig. 2 has four reference lines, at which the strip degrees of freedom are located, and this corresponds to cubic representation. However, the fields differ radically between the two approaches in the longitudinal representation of displacement quantities.

In the S-a FSM the longitudinal representation of each of the five quantities is by a multi-term series of r analytical functions which are continuous over the full structure length [2,4,5]. It is difficult to cope with structures (as distinct from single rectangular plates) which have end conditions other than diaphragm supports (with v = w = ψ_y = 0) and thus it is only for these useful conditions that the multi-term S-a FSM has been developed to a significant extent. Then the longitudinal functions are sine (for v, w and ψ_y) or cosine (for u and ψ_x) series, whose amplitudes are unknown generalised displacement coefficients.

In the B-s FSM [5-9] the continuous analytical functions of the S-a FSM are replaced with B-spline polynomial functions. The longitudinal B-spline functions can correspond to different degrees of polynomial representation and use has been made of linear, quadratic, cubic, quartic and quintic B-splines, which are designated as B_k-splines, with k = 1, 2, 3, 4 and 5. In using the spline functions the length A is divided into q spline sections which are commonly of equal length d as shown in Fig. 3(a). Corresponding to the q sections there are q + 1 spline knots within the length A, plus other knots outside each end of the length A which are required to complete the definition of a function and to prescribe appropriate end conditions. As an example, Fig. 3(b) shows a local B_3-spline function and Fig. 3(c) shows the combination of local functions which contribute to the complete variation of each of the displacement quantities along a strip. Each local function extends only over a certain number of sections, e.g. a local B_3-spline function extends over four sections, and is associated with an unknown displacement coefficient. A particular point to note is that ψ_x is represented by B-splines of degree k-1 whilst the other four displacement quantities are represented by B-splines of degree k. This corresponds to the so-called $B_{k, k-1}$-spline approach and is introduced to avoid the detrimental effects of shear-locking behaviour.

The spline FSM is much more versatile than is the S-a FSM with regard to the structure end conditions that can be accommodated. However, the S-a FSM is usually the more efficient procedure for structures with diaphragm ends, so long as significant anisotropy is not present.

2.3.2 Single-term analysis of "long" structures
There are situations in which the buckled mode shape of a prismatic structure can reasonably be assumed to be purely sinusoidal in the longitudinal direction. Then it is only necessary to consider behaviour over one prescribed half-wavelength using a single term to represent the variation of each displacement quantity in the longitudinal direction [1,3].

In circumstances where the ends of the structure are supported by diaphragms, where the component plates are made of orthotropic material and where no shear stress is applied, the assumption of a sinusoidal mode in the longitudinal direction is perfectly correct. Then the nodal lines of a mode are straight and parallel to the ends of the structure and the analysis simply becomes a special case of the multi-term S-a FSM, with the analysis length taken to be λ, the prescribed half-wavelength, rather than A.

Where one or more of the component plates is made of anisotropic material and/or where applied shear stress is present, the nodal lines are generally curved and skewed across the structure. Within the restriction of a single-term approach this behaviour can be accommodated if the displacement quantities are represented as complex algebra quantities [3]. When this is done the conditions at the ends of the analysis half-wavelength do not equate to conditions that could apply at the ends of a structure of finite length. Thus the single-term complex analysis is strictly correct only for structures of infinite length but will be approximately correct when the structure is much longer than is the half-wavelength, i.e. when the structure is "long" and the mode is local in nature.

2.4 Superstrips and solution
In modelling a plate or shell structure any component plate can be represented by one or more finite strips and the structure matrices $\mathbf{K}$ and $\mathbf{K_g}$ could be assembled using the normal direct stiffness procedure, and then the linear eigenproblem could be solved. However, it is much more efficient to make use of multi-level substructuring techniques, across the structure, to reduce drastically the number of effective degrees of freedom and the solution time (of what is then a nonlinear eigenproblem) [2-9].

Substructuring is first used at the level of each individual finite strip to eliminate the freedoms at all internal reference lines. Then, at the level of a component plate, repetitive substructuring is used whilst creating an assembly of 2^c identical strips (where c = 0, 1, 2....) to represent the component plate by a process of "doubling up". The assembly of 2^c strips is called a superstrip of order c, or simply a SuperstripC [2]. Typically we may take c = 5, i.e. each plate is modelled with one Superstrip5 which is an assembly of 32 strips, with effective freedoms located only at the outside edges of the plate. There is no loss of accuracy in the substructuring procedure, i.e. the performance of a Superstrip5 is precisely the same as that of an assembly of 32 individual strips, and superstrips of high order can be created without any great time penalty as compared to using just one individual strip.

Beyond the superstrip level and following transformation to a global configuration, higher levels of substructuring can be used for repetitive structures, which involve progressive breakdown of the structure into substructures which are assemblies of two or more component plates. The whole multi-level substructuring process makes it possible to solve efficiently problems of very considerable complexity, with hundreds of thousands of freedoms. Eigenvalues, i.e. buckling stresses, are determined using an extended Sturm sequence - bisection approach and eigenvectors can be determined and used in plotting mode shapes in conjunction with a three-dimensional graphics routine.

2.5 Computer software

Computer software packages have been developed to implement the various analysis capabilities. For the multi-term S-a FSM two separate packages, which are very similar in their overall construction, have been created to conduct analyses based in the contexts of shear deformation and classical plate theories. These packages are given the acronyms BAVAMPAS and BAVAMPAC, respectively [2] : they are in fact restricted to flat plate structures, but other (unnamed) programs extend the capability to shell structures. For the single-term complex S-a FSM analysis of plate structures the packages, for shear deformation and classical theories, are BAVPAS and BAVPAC [3]. For the B-s FSM the comprehensive package PASSAS [8,9] embraces the analyses of both plate and shell structures, in the contexts of both shear deformation and classical theories.

2.6 Typical applications

2.6.1 NASA example 6 panel under combined loading

Fig. 4(a) shows an overall view of this complete corrugated panel and its loading whilst Fig. 4(b) gives geometric details of one of the six repeating sections. The panel is simply supported around its boundary and is of square planform with a side length of 762mm. Component flat plates are symmetric graphite-epoxy laminates which are slightly anisotropic in bending and are thin (1.969mm for the horizontal plates and 1.113mm for the inclined plates). The applied loadings are six combinations of longitudinal compressive force N_x° and shearing force N_{xy}° per unit width of panel and buckling results are quoted in terms of a load factor. Full details of this panel and buckling predictions are given in Refs [2,10]. The original results given in Ref. [10] are accurate finite element values, using a fine mesh of 1728 rectangular elements in the context of classical theory.

Numerical results are recorded in Table 1. For the finite strip method, results have been obtained using the multi-term S-a FSM (with r = 7 terms) and using the spline FSM (with up to q = 8 spline sections) in the context of both classical (CPT) and shear deformation (SDPT) plate theories. In modelling the cross-section each component plate flat is represented by a cubic Superstrip5 and further substructuring is used to reduce the effective freedoms to those at only the outside edges and the junctions between the six repeating sections. The two types of finite strip results compare closely with one another and with the comparative finite element results.

Table 1 Buckling results for NASA example 6 panel

Plate theory	Solution method	Applied loads, kN/m					
		0*	87.57	175.13	350.26	875.65	175.13
		175.13*	175.13	175.13	175.13	175.13	0.0
		Load factor					
SDPT	Spline FSM						
	q = 1	1.6719	1.4772	1.2835	0.8322	0.3439	1.7307
	q = 2	1.3705	1.2354	1.0917	0.7235	0.3000	1.5107
	q = 3	1.2941	1.1731	1.0422	0.7093	0.2959	1.4919
	q = 4	1.2519	1.1407	1.0211	0.7066	0.2952	1.4890
	q = 5	1.2365	1.1293	1.0142	0.7059	0.2951	1.4885
	q = 6	1.2316	1.1257	1.0120	0.7057	0.2950	1.4884
	q = 7	1.2299	1.1244	1.0112	0.7057	0.2950	1.4883
	q = 8	1.2291	1.1239	1.0108	0.7056	0.2950	1.4883
	S-a FSM						
	r = 7	1.2333	1.1271	1.0130	0.7058	0.2950	1.4883
CPT	Spline FSM						
	q = 8	1.2422	1.1348	1.0188	0.7062	0.2951	1.4889
	S-a FSM						
	r = 7	1.2461	1.1379	1.0209	0.7063	0.2952	1.4889
	FEM [10]	1.2480	1.1395	1.0223	0.7077	0.2958	1.4918

*Upper value is N_x°, lower value is N_{xy}°

2.6.2 NASA advanced structural panel under compression

The cross-section of this "advanced" structural panel, described in Ref. [11], is shown in Fig. 5(a). The panel is made of isotropic material ($E = 71.016$ GN/m^2, $v = 0.33$), has diaphragm ends and clamped longitudinal edges, and is subjected to a uniform longitudinal stress σ_x°. The buckling response is such that the buckling mode shape is purely sinusoidal along the structure, with m half-waves over the length A, or in other words with a half-wavelength of A/m. Thus, analysis need be conducted only over one half-wavelength and this has been done using both the single-term S-a FSM [1] and the spline FSM using q = 4, and with one Superstrip5 per component plate [8] for a number of prescribed values of A/m. Finite strip results, using the B-s FSM software package PASSAS and showing the variation of buckling stress with half-wavelength are given in Fig. 5(b) together with the earlier results [11], and there is very close comparison between the two sets of results.

2.6.3 Farnborough curved panel under compression

Snell and Greaves [12] have considered the buckling (and postbuckling) of the blade-stiffened curved panel whose actual cross-section is shown in Fig. 6(a). The panel is made of carbon fibre reinforced plastic, with material property and lamination details given in Refs [12, 9]. It has a length of 540mm and the straight longitudinal edges are free. The panel has been tested experimentally under progressive uniform end shortening. (In fact, three nominally identical panels were tested, giving different buckling loads with the most reliable result, reflecting the least influence of initial imperfection, being 107 kN). The actual test panel is assumed to have effectively clamped ends but here FSM predictions [9] are considered for both clamped and simply supported ends.

The FSM model of the cross-section, which is assumed to carry uniform stress σ_x°, is shown in Fig. 6(b), wherein we have that:

lay-up A = [+45°/0°/-45°/0°]$_{2s}$ with thickness 2mm;
lay-up B = [+45°/0°/(+45°/0°/-45°/0°)$_2$]$_s$ with thickness 2.5mm.

The model uses one Superstrip6 to represent each of the eleven component plates. Predicted and measured values of the buckling load are recorded in Table 2 and include a convergence study with respect to q for the spline FSM (using PASSAS) and a value for the S-a FSM with r = 6 (using BAVAMPAS). Close comparison is exhibited between these results and the experimental value. The buckled mode shape of the panel with clamped ends is shown in Fig. 6(c).

Table 2 Values of buckling load (kN) for the Farnborough curved panel

Solution method	Diaphragm ends	Clamped ends	
	TST analysis	TST analysis	SDST analysis
Spline FSM: q = 1	113.8	-	-
q = 2	110.9	156.3	151.0
q = 3	110.6	143.8	134.6
q = 4	110.5	130.7	126.1
q = 5	110.5	124.0	121.9
q = 6	110.5	118.2	116.4
q = 7	110.4	114.8	113.4
q = 8	110.4	114.3	112.9
q = 9	110.4	113.2	111.8
q = 10	110.2	112.3	111.0
q = 11	109.8	111.6	110.3
q = 12	109.6	111.2	109.9
S-a FSM; r = 6	109.8		
Experiment [12]		107.0	107.0

3   OTHER TYPES OF ANALYSIS

3.1   Extended range of buckling applications for laminated structures

Section 2 has been concerned with the buckling of composite laminated plate and shell structures which are single span and which are purely prismatic. However, the capability has been extended recently to include some additional features which allow for some variation of geometry or of support along the length of a structure. This has been done with the spline FSM with the use of a revised longitudinal representation of displacement quantities in which the spline sections can now be of unequal length, thus facilitating the arbitrary positioning of changes along the structure length [13].

The new facility allows for: (a) the presence of rigid point or line supports along the structure; (b) transverse elastic-beam supporting structures; (c) step changes in thickness of component plates along the structure [14]. The developments are made in the contexts of both SDST and TST and again incorporate multi-level substructuring procedures, including superstrips.

3.2   Buckling of sandwich plate structures

The spline FSM analysis described in Section 2 has been extended to structures in which a flat rectangular plate of sandwich construction may, or may not, be reinforced with longitudinal stiffeners [15].   The sandwich comprises two outer facings, typically of laminated construction and perhaps unequal, and an internal core. The facings in effect are modelled as either shear deformable or thin plates and the core is modelled as an orthotropic three-dimensional body.   Thus an increased number of independent displacement quantities is involved through the thickness (8 or 12 for shear deformable or thin facings, respectively). The longitudinal stiffeners are of compact laminated construction. The in-plane stress system is assumed to be applied through the facings.

The buckling response of sandwich plates is complicated by the possibility of highly-localised modes, such as face-wrinkling modes, as well as of overall modes.   The developed FSM capability can accommodate both types of mode in a unified approach [15]. As an example of face-wrinkling behaviour the buckling of a square sandwich plate of side length 228mm with simply supported edges, has been considered. The anisotropic facings are $45°/0°/45°$ CFRP laminates of thickness 0.5mm and the orthotropic core is 25mm thick. The plate is subjected to direct stress in the y-direction and the buckled mode of the top facing is shown in Fig. 7(a). This complicated mode would clearly be difficult to model whatever approach were to be used.   Here it is accommodated by using thirty spline sections in the x-direction and a superstrip order of up to $c = 8$ (i.e. 256 strips across the plate). Table 3 gives details of the convergence of the buckling load with increase in c.   The rate of convergence is much slower than is usual, due to the complexity of the mode shape, but results are effectively converged at $c = 7$. It is noted that a comparative solution exists for this problem [16] of $N_y° = 285$ N/mm, but this omits the anisotropic coefficients, A_{16} etc, from consideration and hence is a significant over-estimation.

Table 3 Buckling results for a sandwich plate with anisotropic facings

Value of c	No of Strips	Values of $N_y°$ (N/mm)	
		CPT facings	SDPT facings
3	8	424.4	421.6
4	16	292.8	277.1
5	32	240.7	232.2
6	64	232.9	228.0
7	128	232.2	227.8
8	256	232.2	227.8

3.3   Thermal buckling
The FSM can be used in predicting a critical temperature change that leads to buckling, whether that change be uniform or non-uniform over the surface [17]. Up to the present the spline FSM has been used in analysing the thermal and thermo-mechanical buckling of single, flat, rectangular laminates. In general it is necessary to conduct an initial plane stress analysis to determine the in-plane stress distribution due to temperature change, before proceeding to the solution of the buckling problem.

3.4   Postbuckling behaviour
The FSM has been used in predicting the postbuckling behaviour of both single rectangular plates and prismatic plate structures. The problem is treated as one of geometric nonlinearity using a total Lagrangian approach in which the nonlinearity is introduced in the strain-displacement equations in the manner of the von Karman assumptions. The nonlinear equilibrium equations are solved using the Newton-Raphson procedure. The basic problem considered is the practically-important one of the response of a plate or plate structure to a progressive uniform end-shortening strain, ε, although other types of problem can also be solved. Developments have been made in the contexts of both CPT and SDPT [18-23].

In considering the response to progressive end shortening of plate structures the S-a FSM has been applied in two quite different ways. In the first approach, post-local-buckling is considered and this is based on the adoption of the "classical" assumptions of post-local-buckling analysis relating to the movements of the junctions between component plates [20]. The adoption of these assumptions, which restricts attention to balanced, orthotropic structures, means that it is possible to base the analysis on a length of the structure equal to one half-wavelength of the initial buckling mode. In the second approach the post-overall-buckling of structures of general lamination, with diaphragm ends, is considered [21]. The analysis is conducted over the whole structure length using extended forms of the von Karman equations, with nonlinear terms in the crosswise displacement v now included.

The results of typical applications of the two S-a FSM approaches are illustrated in Figs. 8 and 9, and in both cases comparison is made with FEM analysis using LUSAS. The first application concerns the post-local-buckling of an orthotropic, five-layer box column of square cross-section [20] (A/B = 3, B/h = 20, where B and h are the breadth and thickness of the component plates, σ_{av} is average longitudinal force and w_{max} is the maximum deflection). The second concerns the overall response to end shortening of a five-layer orthotropic plate stiffened with two longitudinal blade stiffeners located at the quarter points across the plate [21]. (A = 200cm, B (plate width) = 80cm, h_1 (main plate thickness) = 0.8cm, d (blade height) = 8cm, h_2 (blade thickness) = 0.4cm, F is the total end thrust and w is the central deflection.) Close comparison is exhibited between the S-a FSM and FEM results.

The spline FSM can also be used in post-buckling analysis, of course, and, as in buckling analysis, is intrinsically more versatile than is the S-a FSM. Application of the spline FSM to the postbuckling analysis of single plates is described in Refs [22,23]. A similar, enhanced capability for prismatic plate structures has also been developed and used successfully in the solution of complex problems [24].

The work conducted to date on postbuckling analysis using the FSM concerns flat plates and plate structures. This could be extended to embrace curved shell structures.

3.5 Dynamic analysis

Although it falls outside the remit of the present paper it bears mentioning that the FSM has also found considerable usage in the solution of dynamic problems.

So far as natural frequency calculation is concerned, the procedure is very closely related to that for the calculation of buckling stresses, simply with the consistent mass matrix replacing the geometric stiffness matrix (or both being present for problems involving vibration in the presence of initial stress). All the buckling work described in Section 2 has its counterpart in natural frequency calculation and the software packages noted in Section 2.5 all embrace frequency calculation. The same close correspondence between the calculation of buckling stresses and natural frequencies also applies for sandwich plate structures (Section 3.2).

The FSM has also been used to some extent in predicting transient response to dynamic loading. For laminated construction this use has to date been restricted to single plates [25,26] but extension to plate and shell structures is clearly possible.

4   CONCLUDING REMARKS

Description has been given of the use of the FSM in predicting the buckling and postbuckling behaviour of stiffened plate and shell structures. For composite laminated structures, analyses are based on the use of both thin, or classical, theory and of first-order, shear deformation theory, whilst for sandwich plates a higher-order theory is used.

For complicated prismatic laminated structures the FSM is an attractive and powerful procedure for predicting buckling stresses (and natural frequencies) as part of the design process and its efficiency is generally superior to alternative procedures, often markedly so. The FSM is also a very useful tool for predicting nonlinear postbuckling response and its use in this area is developing.

REFERENCES

1.   Dawe, D.J., Finite strip buckling analysis of curved plate assemblies under biaxial loading. *Int. J. Solids Struct.*, **13**, 1141-55, 1977
2.   Dawe, D.J. and Peshkam, V., Buckling and vibration of finite-length composite prismatic plate structures with diaphragm ends, parts I and II. *Computer Meth. Appl. Mech. Engng*, **77**, 1-30 and 227-52, 1989
3.   Dawe, D.J. and Peshkam, V., Buckling and vibration of long plate structures by complex finite strip methods. *Int. J. Mech. Sci.*, **32**, 743-66, 1990
4.   Mohd, S. and Dawe, D.J., Buckling and vibration of thin laminated composite, prismatic shell structures. *Composite Struct.*, **25**, 353-62, 1993
5.   Dawe, D.J., Finite strip buckling and post-buckling analysis. In *Buckling and Postbuckling of Composite Plates*, (Ed. G.J. Turvey and I.H. Marshall), Chapman & Hall, 1995, pp 108-53

6. Dawe, D.J. and Wang, S., Buckling of composite plates and plate structures using the spline finite strip method. *Composites Engng*, **4**, 1099-1117, 1994

7. Wang, S. and Dawe, D.J., Spline finite strip analysis of the buckling and vibration of composite prismatic plate structures. *Int. J. Mech. Sci.*, **39**, 1161-80, 1997

8. Dawe, D.J. and Wang, S., Buckling and vibration analysis of composite plate and shell structures using the PASSAS software package. *Composite Struct.*, **38**, 541-51, 1997

9. Wang, S. and Dawe, D.J., Buckling of composite shell structures using the spline finite strip method. *Composites Part B*, **30**, 351-64, 1999

10. Stroud, W.J., Greene, W.H. and Anderson, M.S., Buckling loads of stiffened panels subjected to combined longitudinal compression and shear : results obtained with PASCO, EAL and STAGS computer programs. *NASA* TP2215, 1984

11. Viswanathan, A.V. and Tamekuni, M., Elastic buckling analysis for composite stiffened panels and other structures subjected to biaxial inplane loads. *NASA* CR-2216, 1973

12. Snell, M.B. and Greaves, L.J., Buckling and strength characteristics of some CFRP stiffened curved panels. *Thin-walled Struct.*, **11**, 149-76, 1991

13. Tan, D. and Dawe, D.J., General spline finite strip analysis for buckling and vibration of prismatic composite laminated plate and shell structures. *Composite Part B*, **29B**, 377-89, 1998

14. Dawe, D.J. and Tan, D., Finite strip buckling and free vibration analysis of stepped rectangular composite laminated plates. *Int.J Num. Meth. Engng*, **46**, 1313-1334, 1999

15. Yuan, W.X., Buckling and vibration of compact laminated and sandwich plate structures by a refined spline finite strip analysis. PhD thesis, The University of Birmingham, 1999

16. Pearce, T.R. and Webber, J.P.H., Buckling of sandwich panels with laminated face plates. *Aero. Quart.*, **23**, 148-60, 1972

17. Dawe, D.J. and Ge, Y.S., Thermal buckling of shear-deformable composite laminated plates by the spline finite strip method. *Computer Meth. Appl. Mech. Engng.* (To appear)

18. Dawe, D.J., Lam, S.S.E. and Azizian, Z.G., Nonlinear finite strip analysis of rectangular laminates under end shortening, using classical plate theory. *Int. J. Num. Meth. Engng.*, **35**, 1087-110, 1992

19. Lam, S.S.E., Dawe, D.J. and Azizian, Z.G., Nonlinear analysis of rectangular laminates under end shortening, using shear deformation plate theory. *Int. J. Num. Meth. Engng.*, **36**, 1045-64, 1993

20. Dawe, D.J., Lam, S.S.E. and Azizian, Z.G., Finite strip post-local-buckling analysis of composite prismatic plate structures. *Comput. Struct.*, **48**, 1011-23, 1993

21. Wang, S. and Dawe, D.J., Finite strip large deflection and post-overall-buckling analysis of diaphragm-supported plate structures. *Comput. Struct.*, **61**, 155-70, 1996

22. Dawe, D.J. and Wang, S., Postbuckling analysis of thin rectangular laminated plates by spline FSM. *Thin-walled Struct.*, **30**, 159-79, 1998

23. Wang, S. and Dawe, D.J., Spline FSM postbuckling analysis of shear-deformable rectangular laminates. *Thin-walled Struct.,* **34**, 163-178, 1999

24. Dawe , D.J. and Wang, S., Postbuckling analysis of composite laminated panels. *AIAA J.* (to appear)

25. Chen, J. and Dawe, D.J., Linear transient analysis of rectangular laminated plates by a finite strip - mode superposition method. *Composite Struct.*, **35**, 213-28, 1996

26. Wang, S., Chen, J. and Dawe, D.J., Linear transient analysis of rectangular laminates using spline finite strips. *Composite Struct.*, **41**, 57-66, 1998

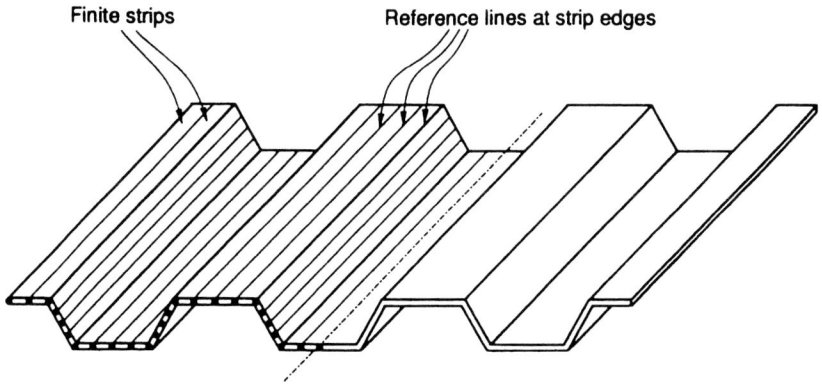

Fig. 1 Typical plate structure modelled with finite strips

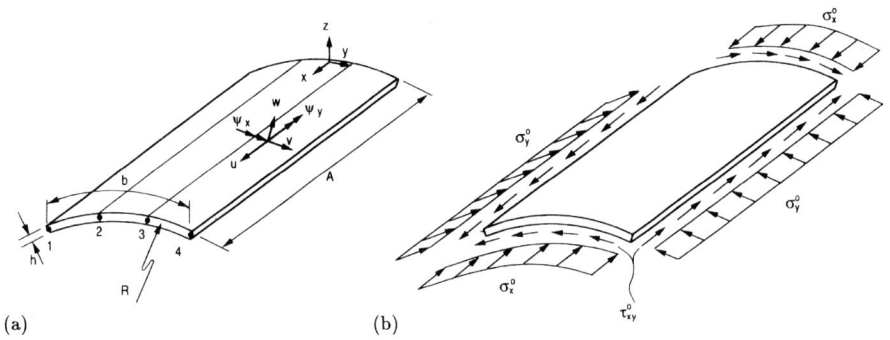

(a)   (b)

Fig. 2 A curved finite strip: (a) geometry and displacements;
(b) applied stress system

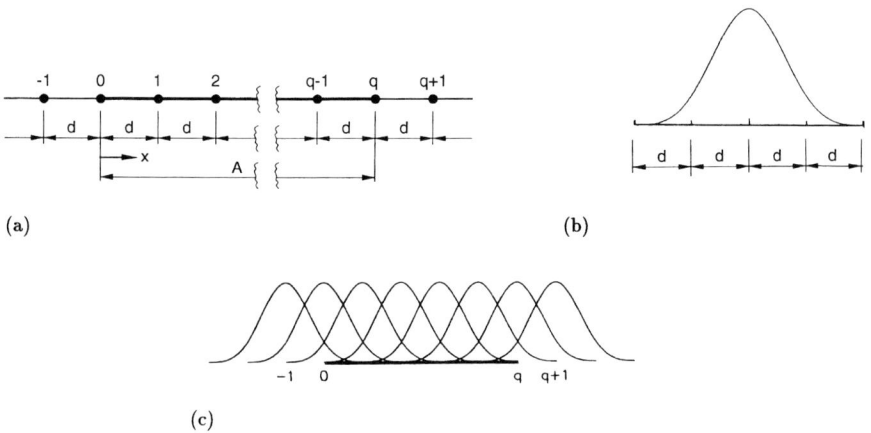

(a)   (b)

(c)

Fig. 3 Spline representation: (a) spline sections and knots; (b) local cubic
spline function; (c) a combination of local cubic spline functions

                                   S695/002/99 © IMechE 2000

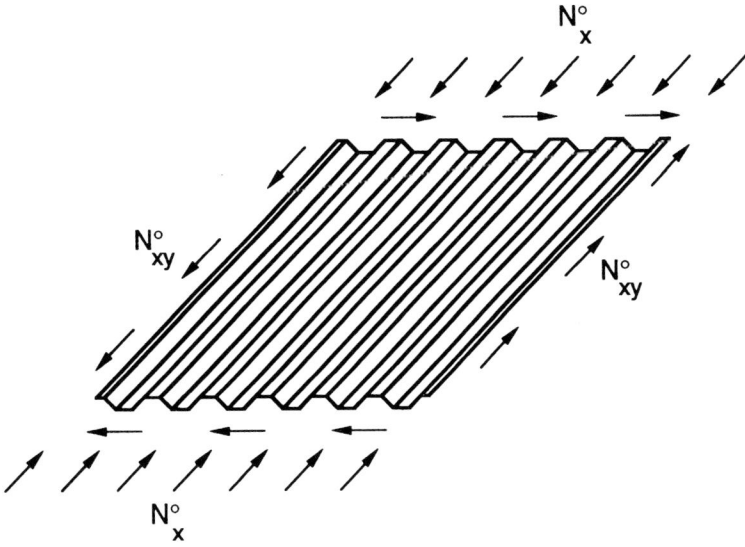

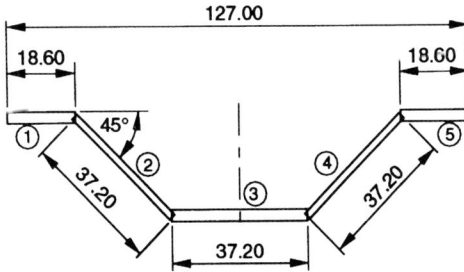

Fig. 4 The NASA Example 6 panel: (a) general view; (b) details of a repeating element (with dimensions in mm)

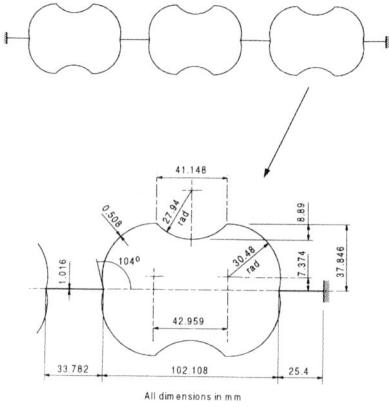

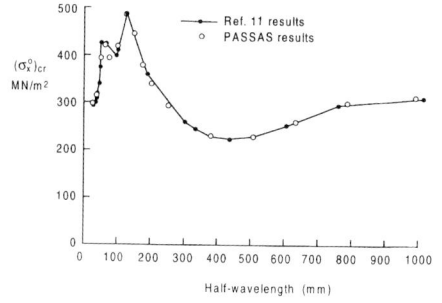

(a)

(b)

Fig. 5 The NASA advanced structural panel: (a) panel cross-section;
(b) buckling stress versus half-wavelength

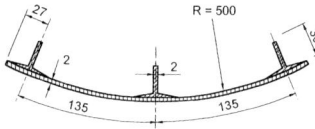

(a)

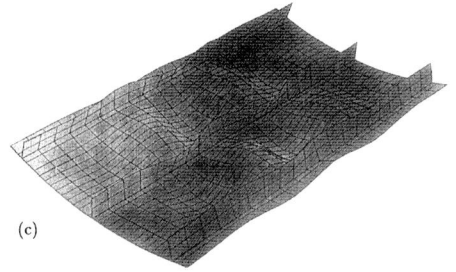

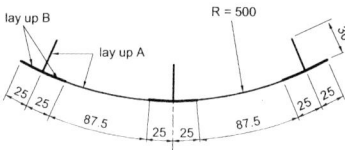

(b)

(c)

Fig. 6 The Farnborough curved panel: (a) actual cross-section;
(b) modelled cross-section; (c) buckled mode of panel with clamped ends

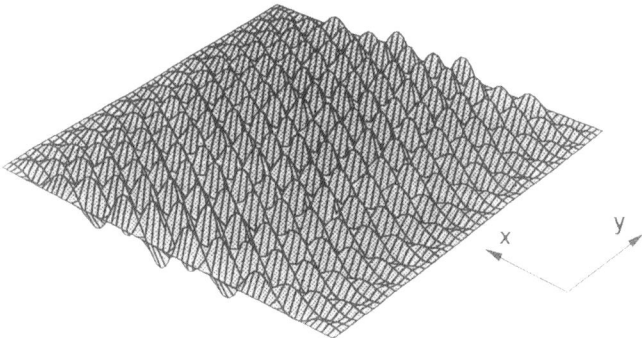

Fig. 7 Buckled mode shape of a sandwich plate with anisotropic facings

S695/002/99 © IMechE 2000

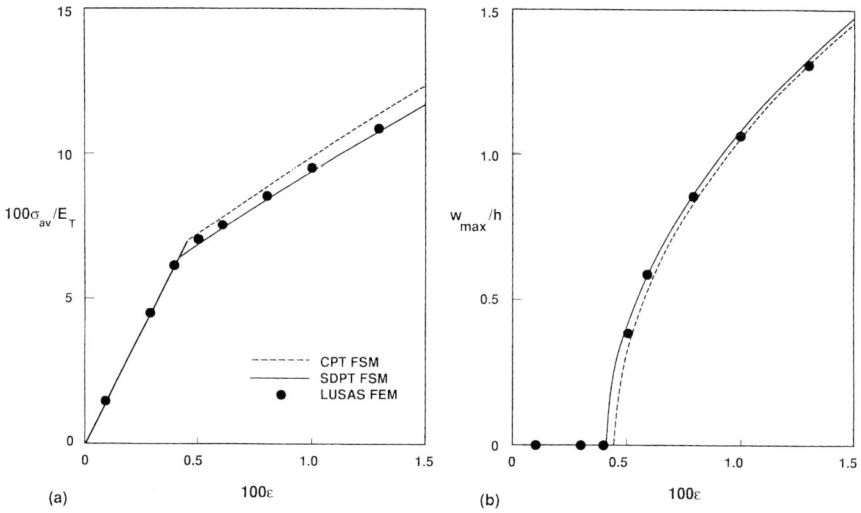

Fig. 8 Nonlinear response of orthotropic box to applied shortening strain: variation of (a) average longitudinal stress, and (b) maximum deflection

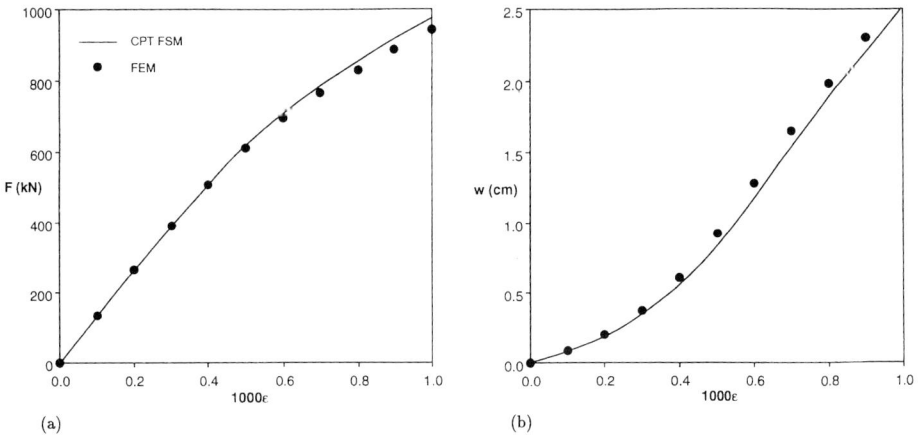

Fig. 9 Nonlinear response of blade-stiffened panel to applied shortening strain: variation of (a) end thrust, and (b) central deflection

S695/003/99

Design and analysis of stiffened composite structures at DERA

P HOPGOOD, S SINGH, and E S GREENHALGH
Mechanical Sciences Sector, DERA, Farnborough, UK

ABSTRACT

This paper reviews work on stiffened structures that has been undertaken within the Structural Materials Centre at DERA Farnborough. The activities described are mainly concerned with applications in polymer-composite aircraft structures, and stringer-stiffened panels in particular. The paper presents a brief overview of the advantages and disadvantages of some of the stiffening methods available. The importance of damage tolerance is discussed, and key mechanisms governing failure due to compression after impact are described.

INTRODUCTION

Stiffening is used to stabilise thin-skinned structures and panels, and is particularly important for the prevention of buckling under compressive loading. Stiffening can be achieved in a number of ways. The simplest methods are to thicken the structure, or to achieve a degree of geometric stiffening by curving the structure. For high-performance structures, these approaches are generally either weight-inefficient or impractical. In practice, sandwich constructions, or the use of discrete stiffeners are usually the preferred solutions.

Sandwich construction is popular for relatively lightly loaded, thin-skinned panels, and for relatively thin panels that can utilise the full-depth core efficiently. Sandwich construction is also favoured for stiffening panels that have complex geometry. In such cases, the use of discrete stiffeners can be costly, impractical, and may give less control over the surface geometry. Sandwich construction is also attractive for applications in which both faces of the panel must be flat, and, because of their good omni-directional performance, for panels that are subjected to a variety of multi-axial loading conditions. However, sandwich panels have become unpopular with some users due to susceptibility to damage, difficulty of repair and the tendency to entrap water. This has led some manufacturers to reconsider the use of sandwich construction, and to look again at using discrete stiffening even for relatively lightly loaded structure. Some have started to revert to metallic structure, especially for parts of the aircraft particularly prone to impacts and other forms of mechanical abuse. It is the contention of the authors that this last step represents something of an over-reaction and that for many applications discrete stiffened composite construction is the most suitable choice.

In current practice, discrete stiffening is generally preferred over sandwich construction for more highly loaded, thicker-skinned structures, and for structures that are subjected to well defined loading conditions to which uni-axial, or orthogonal stiffening is appropriate. Discrete stiffening with stringers is common in primary aircraft structures in both the fuselage and wing areas. For wings, discrete stiffeners offer the additional advantage that the volume between the stringers is available for fuel storage.

Activities at DERA have been concerned with impact damage and residual strength of both sandwich panels and stringer-stiffened panels. These have included:

- Prediction of the onset of buckling using both finite-element and finite-strip techniques
- Experimental studies of residual strength after impact on thin stringer-stiffened panels for lightweight aero-structures. This included dynamic finite-element modelling of damage introduction during impact, (UK DTI DSIONS Programme - Hurel Dubois / DERA / Hexcel)
- Impact and residual strength of stringer-stiffened panels (M.O.D. Applied Research, DAMOCLES , E.C. EDAVCOS)
- Identification of delamination growth mechanisms in plain and stringer-stiffened panels (MOD Applied Research, DAMOCLES)
- Finite-element modelling of delamination growth in plain and stringer-stiffened panels (MOD Applied Research)
- Impact and residual strength of sandwich panels (EC Brite Euram DAMTOS)
- Evaluation of novel stiffening methods using z-pinned cores and 3D fibre pre-forms

Recently, most work has concerned the damage resistance and tolerance of stringer-stiffened panels, and it is this work which is reported here.

WHY IS DAMAGE TOLERANCE IMPORTANT TO STIFFENED STRUCTURES?

Most structures are susceptible to impact threats from, for example, dropped tools, hail, or runway debris. In carbon-fibre composites, damage introduced by these impacts may not be visible to the naked eye. Certification procedures dictate that an allowance must be made for the presence of a level of damage that cannot be detected by a visual inspection. Such invisible damage is most likely to grow under compressive loading, where additional stresses can be induced by geometric and material instabilities. Stiffened structures are often used in the presence of compressive loading, and therefore achieving damage tolerance in stiffened composite structures is very important.

WHAT ARE THE THREATS?

Aircraft structures are exposed to a range of impact threats that can create barely visible impact damage. Barely visible damage is usually defined by the minimum dent depth that is visible to the naked eye during inspection. There are other threats, such as ballistic impact, lightning strike, bird strike, and collision but these are likely to leave clearly visible damage.

Demonstration of damage tolerance during certification is conventionally based upon defining realistic impact energy levels, and upon residual strength tests. For thin laminates the maximum tolerable impact energy is defined as the highest level that causes barely visible damage. For thick laminates, the maximum impact energy is defined by the highest energy threat, which may be the greatest energy expected from a tool drop [1]. This has been conventionally taken to be 50J within European certification programmes, although levels of 140J are considered in the USA and in some European programmes.

Impact tests are usually undertaken with impactors of a relatively small radius, however it has been shown that objects with larger radii will create a similar level of damage for the same impact energy, but with a much smaller dent depth. It follows that in composites, dent depth is not necessarily a very good guide to the level of damage inflicted.

RESPONSE

Recent work within the DAMOCLES programme [2] has highlighted limitations in the current acceptance levels of damage tolerance. Impacts of small objects at higher speeds have been shown to generate more damage than impacts of heavier objects at lower speeds, even though the energies are equivalent [3,4]. Three distinct regimes of impact response have been identified (Figure 1). The first occurs for thick structures, where the impact times are very short, and the response is dominated by dilatational waves. The second occurs where the mass of the impactor is low; approximately 1/5th of the affected mass of the panel or less. Here, the response is determined by the formation of flexural waves at the impact site, and the response can be independent of the boundary conditions of the panel and the surrounding structure. The third occurs when the mass of the impactor is larger than the mass of the affected panel area. Here, quasi-static loading can be assumed, and the boundary conditions of the panel and surrounding structure are important.

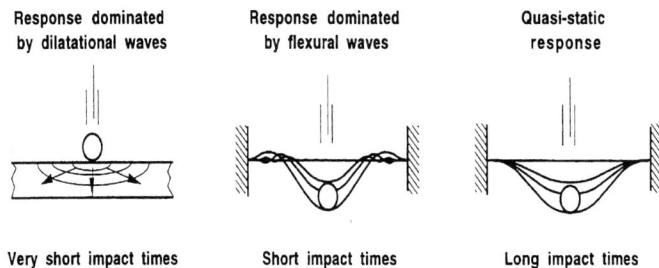

Response dominated by dilatational waves	Response dominated by flexural waves	Quasi-static response
Very short impact times	Short impact times	Long impact times

Figure 1 Three regimes of impact response
(courtesy of R Olsson [5])

A useful model for determining a threshold of impact force, below which delaminations will not be induced, has been developed by Davies at Imperial College [6]. For isotropic laminates, the critical force is given as:

$$P_c^2 = \frac{8\pi^2 E(2t)^3 G_c}{9(1-\upsilon^2)}$$

where E is the Young's Modulus, t is the laminate thickness, G_c is the critical strain-energy release rate, and v is the Poisson's ratio. Using typical data for carbon-fibre reinforced epoxy (E=60GPa, G_c=0.8N/mm, v=0.3) the equation becomes:

$$P_c^2 = 680t^{3/2}$$

This model is suitable for the problems in which the quasi-static loading approach is appropriate. Although the model is simple, it has been found to be surprisingly effective. The model has been used in combination with finite-element analysis to optimise the damage tolerance of stringer-stiffened panels. These studies were undertaken by NLR within the DAMOCLES programme.

PREDICTION OF BUCKLING AND POST-BUCKLING

Buckling predictions have been undertaken using both finite-element and finite-strip techniques. The predictions were used in the design of test panels for battle-damage repair, and for the damage tolerance of stringer-stiffened panels. Suitable accuracy can be achieved with either method, but this can be achieved more readily with a higher solution speed using the finite-strip technique. However, increases in solution speed can often be outweighed by the time taken for engineers to generate the model. The PASSAS finite-strip software, developed at Birmingham University, has been used by DERA to predict the buckling performance of both curved and flat stiffened-panels. The extension of the method into post-buckling has been useful, the savings in solution time have been even more apparent, and the predictions have agreed very well with previous experiments (Figure 2).

Finite strip analysis is applicable to prismatic plate and shell panels, but finite-element analysis is more suited to more complex structures, and for panels containing discontinuities such as damage zones.

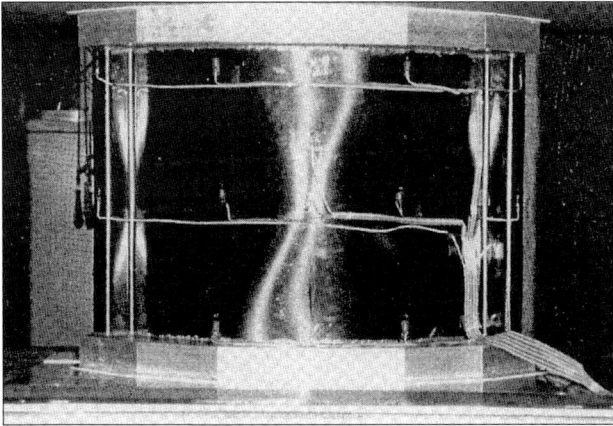

Figure 2 Post-buckled shape of curved battle-damage repair panel correctly predicted by PASSAS finite-strip analysis

THIN STRINGER-STIFFENED PANELS

Work in this area was undertaken within the DSIONS programme that was sponsored by the UK DTI and was led by Hurel-Dubois UK Ltd. The work was related to damage tolerance of lightweight aero-structures. A range of structural design and material technologies were applied to uni-axial stiffened panels (Figure 3). DERA were involved in the design and test of structural elements, and in modelling the introduction of impact damage using dynamic finite-element analysis.

The impact response and the design philosophies for thin panels were found to be considerably different from those for thicker panels. For example, critical dent-depths corresponding to the definition of BVID constitute penetration for very thin panels. This indicated that for thin panels it is necessary to be able to withstand considerable damage including penetration without failing and without the need for major repair. This led to the need for the stiffeners to provide some redundancy. For these and other reasons it was concluded that testing at the element level, rather than with coupons, was vital in obtaining a cost-effective route to the certification of thin stringer-stiffened panels. Considerable progress was made and success achieved in obtaining effective damage tolerance in thin stringer-stiffened panels.

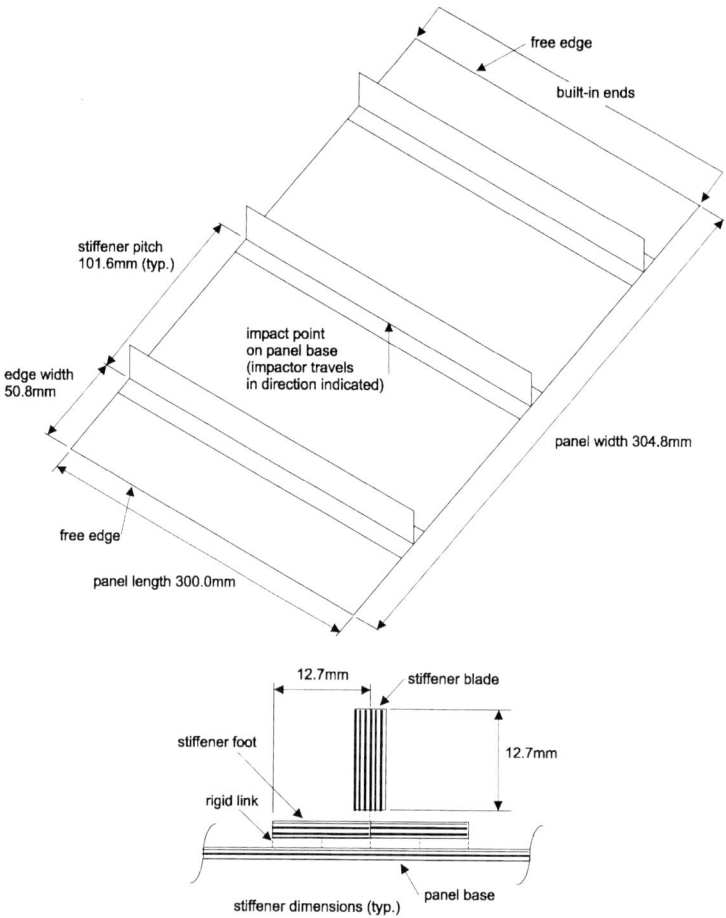

Figure 3 Thin blade-stiffened panels used in the DSIONS programme

Figure 4 illustrates a prediction of damage within a thin stringer-stiffened panel that was made using dynamic finite-element analysis. This work used the FE77 finite-element code, which was developed at Imperial College. The Chang-Chang material degradation model was used to progressively modify the properties of the structure as the impact event progressed. For the limited range of configurations that were modelled, there was good correlation between theory and experiment. The cracking that was predicted within the central stiffener (Figure 4) was also evident when the panel was impact tested.

The next step for this work is expected to be an extension of the testing to 3-dimensional orthogonal structures under multi-axial loading, using DERA's bi-axial test machines.

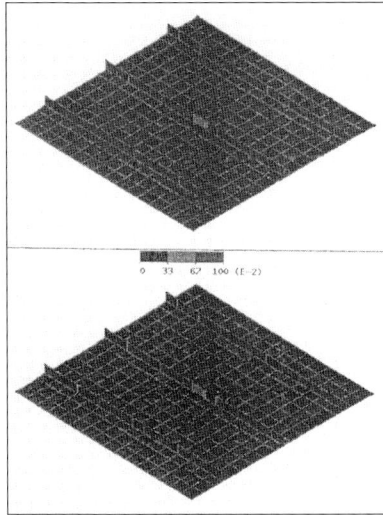

Figure 4 FE77 damage prediction for impact over the central stiffener showing fibre-fracture in the cracked stiffener

THICKER STRINGER-STIFFENED PANELS

Introduction

Recently, the majority of DERA's work on stiffened-panels was concerned with I-section stringer-stiffened panels that were representative of military aircraft structure [7]. Three types of panel were tested, their geometries being illustrated in Figure 5. The type-1 panel had a 4mm thick skin and narrow bays between the stiffeners (small stiffener pitch). The type-2 panel had a 4mm thick skin and wider bays. The type-3 panel had a 3mm thick skin and narrow bays. In all cases the I-stiffeners were 3mm thick, and were co-cured with the skin. The panels were subjected low-velocity impacts of 15J, which represented a perceived threat (Figure 5). It should be noted that this was considerably less than the 50J global threat that is generally applied.

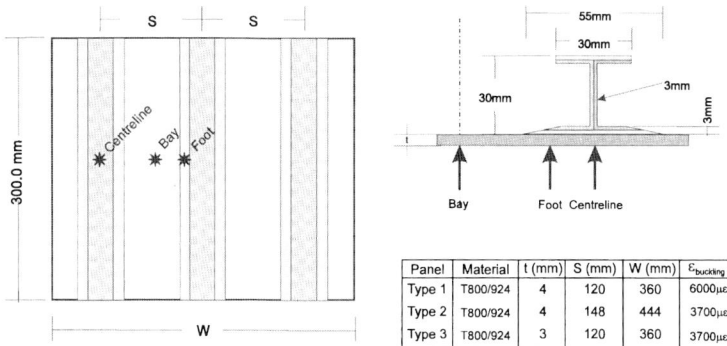

Panel	Material	t (mm)	S (mm)	W (mm)	$\varepsilon_{buckling}$
Type 1	T800/924	4	120	360	6000$\mu\varepsilon$
Type 2	T800/924	4	148	444	3700$\mu\varepsilon$
Type 3	T800/924	3	120	360	3700$\mu\varepsilon$

Figure 5 Dimensions of I-stiffened panels, impact locations, and undamaged buckling strains

The regions of the panels that were most sensitive to impact were identified in two-stages. First, the amount of damage introduced at the various locations was identified using measurements of dent-depth, and damage area. Then the residual compressive strengths of the panels were measured.

The mechanisms of damage introduction and failure were investigated using fractographic techniques, and delamination was found to be a key failure mechanism. Using knowledge of these mechanisms, design guidelines have been developed which can help to minimise impact damage and to maximise residual strength. The work is continuing with design and testing of improved panel designs. Special finite-element modelling techniques, developed at FFA, are also being used to model delamination growth and residual strength of stringer-stiffened panels.

Impacts on I-stiffened panels
In the low velocity impact of carbon-fibre composites, transverse cracking, delamination, and fibre-fracture are the three dominant types of fracture. These fracture mechanisms each absorb impact energy. Energy is also absorbed through the elastic response of the structure. The panel geometry, end constraints, and the location of the impact with respect to the local sub-structure control the degree to which the different mechanisms contribute to the total absorbed energy.

During impact, transverse cracks usually occur first, and these act as initiation sites for delamination. Delaminations can then be driven by the through-thickness shear stresses that are generated as the impactor deforms the panel. Delaminations occur preferentially at orthogonal ply interfaces ($+45^{\circ}$ /-45° or 0°/90°) where there are significant changes in stiffness. Fibre fractures may also occur, and these absorb large amounts of energy. Fibre fractures are driven by the high through-thickness stresses that are generated directly below the impactor. Such fractures can have the most significant effect on residual strength, since the load bearing plies can be considerably weakened.

For the 4mm thick panels with narrow bays, the damage areas were found to be larger for impacts in the bay than for impacts over the stiffener. It was believed that although the impact forces in the compliant bay were lower than those at the stiffener, the combination of forces and high deflections in the bay led to significant delamination. For impacts over the stiffener foot, more energy was absorbed in the elastic response of the stiffener, and damage areas were smaller than for impacts in the bay. For impacts at the stiffener centreline, almost all of the impact energy was absorbed by the elastic deformation of the stiffener, and damage was minimal.

As the width of the bay increased, damage areas for impacts in the bay decreased. This was believed to be due to the increased elastic energy absorption of the wider bays. However, damage areas increased for impacts over the stiffener foot, with the damage extending along the stiffener length. The larger deflections arising from the more compliant bays may have given rise to greater driving forces for delaminations along the stringer length.

For the thinner 3mm thick panels, there was a further energy absorbing mechanism that can be attributed to the change in skin thickness. The skin in this panel type experienced higher deflections and consequently higher interlaminar shears than in the other panel types. This led to a greater density of transverse cracks than observed in the thicker skinned panels. These cracks initiated a higher number of delaminations, thus requiring a smaller damage area to absorb the same amount of incident energy.

For the impacts on the stringer centreline, there were no clear trends as the bay width increased. Almost all the damage was absorbed through structural response and the effect of the bay width was negligible.

The dent depths that were observed were very small, typically up to 0.06mm for the 4mm thick skins, and up to .23mm for the 3mm thick skins. These were smaller than the thresholds for barely visible impact damage (BVID), and structures containing such damage would be required to support ultimate load (check).

Residual compressive strength of I-stiffened panels
The residual compressive strengths of the 3 different panel types that were tested are shown in Figure 6.

Panel	Bay Width	Skin Thickness	Design Strain
Type 1	120mm	4mm	$6000\mu\varepsilon$
Type 2	144mm	4mm	$3700\mu\varepsilon$
Type 3	120mm	3mm	$3700\mu\varepsilon$

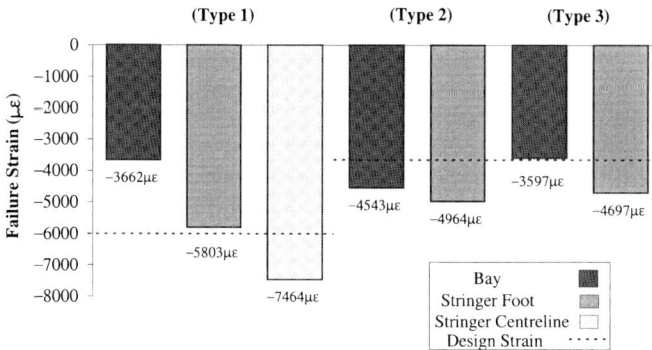

Figure 6 Residual strengths of I-stiffened panels

The overriding factor governing the failure mechanisms was the location of the impact site; the failure processes associated with impact damage in the bay were very different from those arising from impact damage beneath the stiffener feet.

The failure process for a panel containing impact damage in the bay is shown in Figure 7. The first event was the deflection of the impacted bay outwards, away from the stringer face (Figure 7a). The damaged region also started to deflect, but inwards, towards the stringer face (Figure 7b) and the damage lobes started to grow from the impact site (Figure 7c).

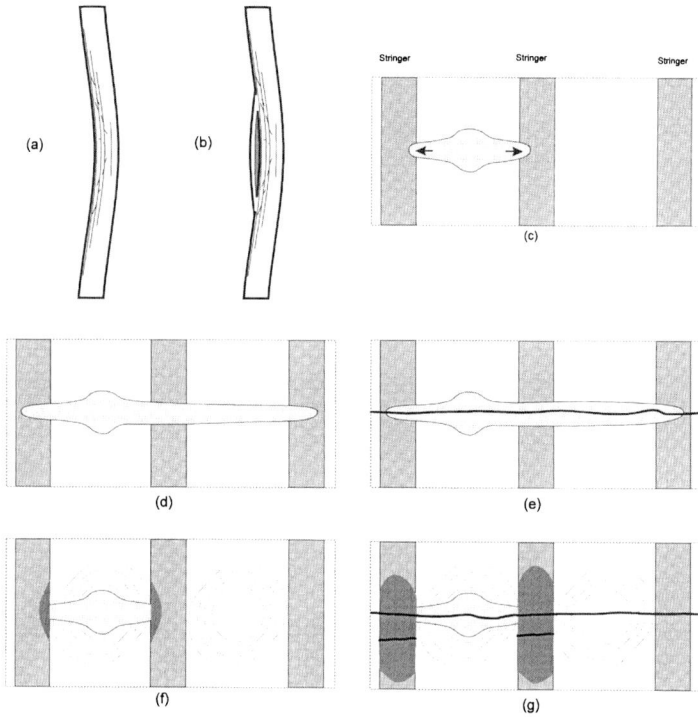

Figure 7 Failure mechanisms for impact damage in the bay

For the Type 1 panel, with 4mm thick skin and narrow bays, the damage growth was stable initially, but reached a point where it became unstable. It then grew within the skin, past the stiffeners, until the panel failed in compression (Figure 7d, 7e). For the other panels, which had lower design buckling strains, the presence of the damage promoted early buckling of the bays. The transverse growth of the damage was arrested at the stiffeners (Figure 7f). As the bays buckled, opening forces came into operation beneath the stiffeners close to the impact site, leading to local detachment of the stiffeners. This loss of local support led to crippling of the skin and thence to compressive failure (Figure 7g).

Damage growth was slower as the bay width increased, leading to a slightly higher strength for the type-2 panel. This was attributed to the out-of-plane deflection of the damaged region being less than that for the other panel types, and the impact damage being more distant from the stiffener feet. This led to two factors that improved the damage tolerance of the design. Firstly, the reduced damage growth led to the damage having less of an influence on buckling performance. Secondly, the damage had less effect on the stiffeners, and did not promote detachment to the same extent as had been seen in the type-3 panel design.

The failure process for the panels containing impact damage over the stiffener foot is shown in Figure 8. The damage growth was less severe than in the panels with damage in the bay, and higher residual strengths were achieved. First, the bay adjacent to the impacted stringer started to deflect outwards (Figure 8a). Then there was evidence of local delamination beneath the impacted stringer foot (Figure 8b). There was no transverse damage growth, but damage may have spread axially, along the stiffener foot. The damage did not grow to a critical level before the undamaged buckling strain of the panels was reached (Figure 8c). It was therefore the design buckling strain that limited the strength of the panels. Buckling of the bays promoted damage development beneath the stiffeners, and led to stiffener detachment. This led to skin crippling, compressive failure initiating from the impact site, and failure of the stringers (Figure 8d).

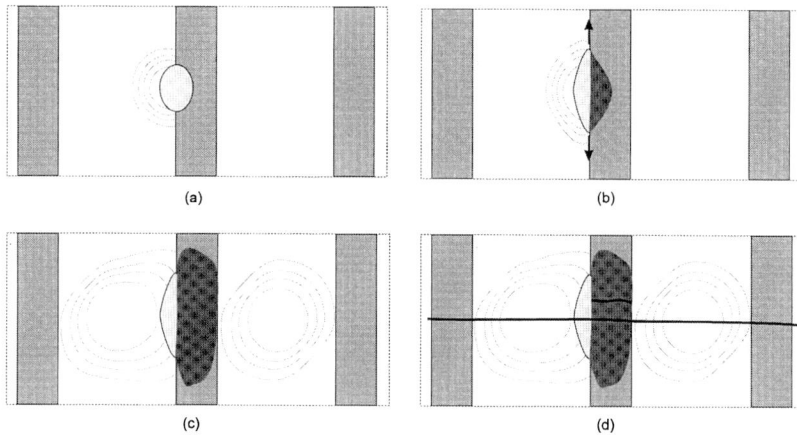

(a)

(b)

(c)

(d)

Figure 8 Failure mechanisms for impact damage in the stiffener foot

Failures of the panels that were impacted over the stiffener centreline were again determined by the undamaged buckling strain of the panel.

Discussion of I-stiffened panel results

The residual compressive strengths of the panels tested here were all relatively low. This was somewhat disturbing in view of the fact that the impact energy used was only 15J as opposed to the target figure of 50J. However, it should be noted that the panels had certain characteristics that made them sensitive to impact damage and to early failure. These characteristics included: co-curing, lay-up, and low buckling strain level. Many panels buckled below ultimate load and failure followed immediately. More representative work has been undertaken on impact of a multi-spar wing-box demonstrator [8]; where impacts of up to 130J were applied, with final failure occurring at approximately 3900$\mu\varepsilon$.

The current design practice for such panels is to design for buckling beyond ultimate load. This means that damage-driving forces that are enhanced by buckling are less likely to occur. However, the results for the type 1 panel, with wide bays and thick skins, show that delamination growth can occur in panels well before they buckle globally. Local buckling of the delamination creates the mechanisms that drive the damage growth. It is therefore important to be able to predict delamination buckling and growth. For this reason, DERA has been involved in finite-element modelling of delamination growth using the ADINA finite element code.

Damage growth from embedded defects in I-stiffened panels

Models such as those described above, combined with finite element stress analysis, allow structures to be designed to be resistant against a particular impact energy level. However designing structures to be resistant against the most severe (but least likely) threat is not practicable. It is therefore important also to consider impact damage tolerance.

In the compression after impact tests carried out at DERA, delamination growth of some form always preceded structural failure. In order to better understand the significance of delamination growth, a number of tests were carried out in stiffened structures containing single artificial delaminations of a similar extent to those typically seen in the impacted panels, in the bays, under the feet of stringers and under stringer centrelines. Delaminations partially or totally beneath the stringers were prevented from buckling; the bays were therefore the most detrimental location for damage. The initial damage growth from the bays was similar to that in unstiffened panels when supported against out-of-plane deflections [9].

Tests and simulations carried out at the Aeronautical Research Institute of Sweden (FFA) have demonstrated that there can be strong interaction between the local buckling of a delamination and the global buckling of a structure [10]. When a structure begins to buckle globally, the loading on a crack front can often be sufficient to drive a crack forward, even though the applied strain level may be well below that which would cause the delamination to grow if it were on a rigid surface.

Careful study of the mechanisms have showed that, contrary to what might be expected, the depth of a delamination, which governs delamination buckling, is not the single most important parameter controlling damage growth. The orientation of fibres relative to the load directions has a greater influence on the damage growth and ultimate strength of a damaged structure. During delamination growth in structures, a shear component of loading tends to drive delamination cracks towards the surface until they encounter fibres oriented in the preferential growth direction (usually transverse to the applied compression load) [11] (Figure 9).

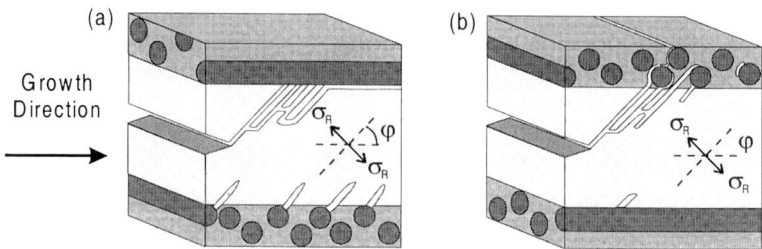

Figure 9 Mixed-mode delamination growth at (a) 0°/φ° ply interface (upper ply oriented in preferential growth direction),and (b) φ°/0° ply interfaces (migration through off-axis ply).

The panels tested at DERA had quasi-isotropic skins of stacking sequence $\{[+45°/-45°/0°/90°]_n\}_s$. With a delamination three plies deep (at the 0°/90° ply interface), the delamination migrated to the –45°/0° ply interface and then the +45°/-45° ply interface but was not able to find any fibres in the 90° direction, transverse to the load, along which to grow (Figures 10-12). With delamination five plies deep at the +45°/-45° ply interface, the crack migrated through one ply toward the surface to the 90°/+45° ply interface. Here the fibres were oriented in the preferred direction (transverse to the load), and growth was rapid (Figure 13). Contrary to conventional concepts, in these panels, the deeper delamination was more detrimental than the shallower one.

The implication of these mechanisms is that when designing damage-tolerant compression panels, the plies nearer the surface (particularly the back-face, which is likely to contain the largest delaminations after an impact) should not have fibres oriented transverse to the applied compression load. A more complete description of the mechanisms and their implications can be found elsewhere [9].

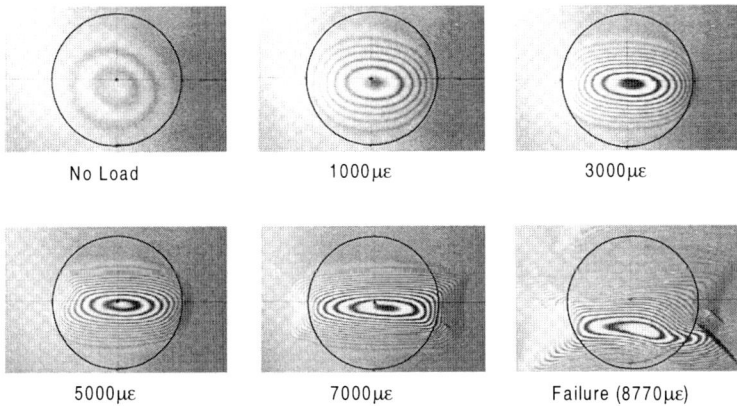

No Load   1000με   3000με

5000με   7000με   Failure (8770με)

Figure 10 Damage growth from 50mm defect three plies deep (0°/90° ply interface)

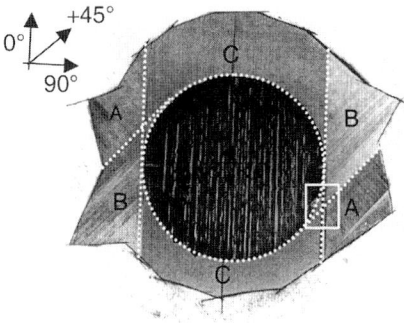

Figure 11: Damage growth from defect three plies deep (0°/90° interface) (lower surface)

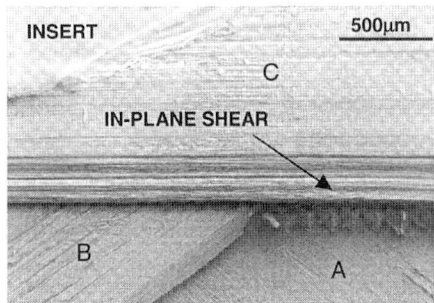

Figure 12: Micrograph of delaminated plies from surface matching Figure 13.

SIMULATING DAMAGE GROWTH IN STIFFENED STRUCTURES

The mechanisms described above are affected by structural buckling mode shapes. It is therefore important to develop models, which can predict the interaction of local buckling and global buckling, and the resulting driving force on a delamination. Such a model, which can additionally simulate a large number of increments of crack growth, has been developed at FFA and is now also being used at DERA. The simulation of delamination buckling and growth is conducted using the commercial code ADINA. The models account for the effects of global bending of the structure and contact between the delaminated plies. The simulation uses a time-stepped series of structural solutions, updated automatically by software developed at FFA.

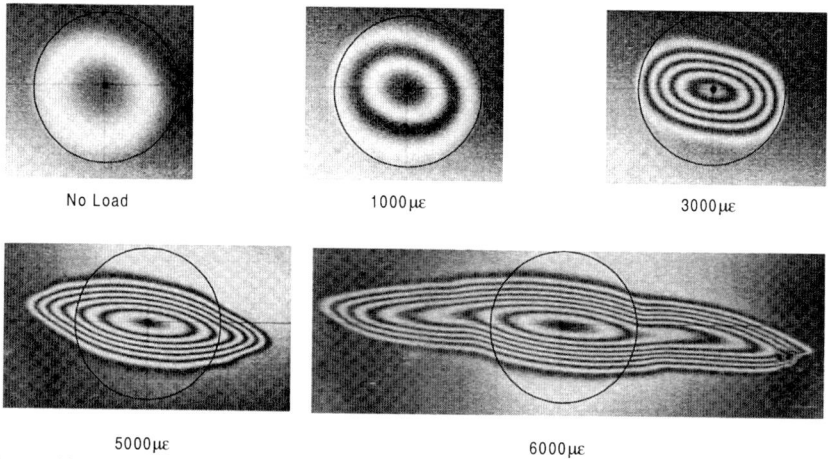

No Load	1000 με	3000 με

5000 με   6000 με

Figure 13 Damage growth from a 50mm defect five plies deep (+45°/-45° ply interface)

One layer of 4-noded mixed interpolation Mindlin/Reissner shell elements represents the delaminated material whilst a second layer represents the sublaminate below. To reduce computation times the constitutive properties of the sublaminate may be represented by homogenised membrane or bending properties. The stacking sequence of the thinner delaminated plies must be represented explicitly, so that stiffness coupling resulting from the unbalanced stack of plies is simulated. Outside the delaminated region, the two layers are joined using nodal multipoint constraint (MPC) equations, ensuring displacement continuity. In the delaminated area, the nodes were coupled in the through-thickness direction only in the event of contact. Separate groups of shell elements represent the foot, web and cap of the stringers. The foot is coupled to the skin with MPC equations, whilst the web is joined to the foot and cap by rigid links. The loaded edges are locked in all degrees of freedom whilst the nodes on the opposing edge are joined by rigid links to a master node (free only to move in the loading direction), to which a point load is applied.

The analysis is performed in the following steps:

- Global buckling analysis (determine when the behaviour became highly non-linear)
- Local buckling analysis of the delaminated plies with contact iteration

- Postbuckling analysis with contact and automatic load increase until the delamination growth criterion (total critical strain energy release rate) was satisfied
- Repeated delamination crack propagation by advancing the mesh in the growth regions

By this approach the delamination propagation is modelled by performing a large number of incremental crack propagations without increasing the number of elements. A detailed description of the moving boundary technique is discussed elsewhere [12]. The moving mesh technique has two distinct advantages over the more common method of simulating crack growth by releasing nodal constraints. Firstly, an initially continuous crack front maintains a continuous profile thereby avoiding the generation of any mesh-dependent spikes in the strain-energy release-rate profile. Secondly, during each growth increment, the connectivities are not changed, and the applied loads and nodal co-ordinates are only slightly modified. Consequently the stiffness matrix is essentially unchanged, and the previous solution can be used as an approximate solution during the time-stepping, avoiding the need to completely restart the analysis after each increment.

The models have been demonstrated to be able to represent the buckling, rotation of the elliptical blister resulting from an unbalanced sublaminate, and the location of the initiation site. It was important that the stacking sequence of delaminated plies was represented explicitly rather than by homogenised orthotropic properties. However, this can generate more complicated buckling mode shapes than those predicted using homogenised properties, which may cause 'snap-through' problems in the analysis.

Although the early stages of damage growth were similar in the plain and stiffened panels, the local substructure had a significant effect in the later stages. As was predicted by the models, the global buckling coupled with the damage, which increased the mode I component at the defect boundary. As the damage approached the stringers, the effect of the stress field changed the behaviour from that observed in the plain panels; the out-of-plane constraint suppressed the mode I component and the growth rate was reduced. For defects partly or completely beneath the stringer feet, this constraint led to massive increases (doubling) in initiation strain.

The models simulate a single delamination of a given size at a known ply interface. In the case of impact damage it is assumed that there exists a sizeable delamination at each interface, and one of these will, due to its orientation and depth, be the most favourable for delamination growth. At present growth from a few alternative ply interfaces may need to be simulated. Analysis times could be reduced if an algorithm were developed to automatically identify this critical ply interface, thus removing the need for the redundant simulations.

These kinds of models will in the future allow optimisation of structures considering damage tolerance as well as damage resistance.

CONCLUSIONS

The successful design and analysis of high-performance stiffened-structures using carbon-fibre composites is highly dependent upon understanding damage and damage tolerance. Considerable progress has been made in this area within DERA, and within DERA's collaborative programmes in Europe. Understanding the mechanisms of failure has enabled design guidelines and design tools to be developed. These have been successfully used to develop new stiffened structures with significantly enhanced damage resistance and tolerance.

There has been particular success in the development of thin stringer-stiffened panels. It was found that structural element tests are particularly important for cost-effective certification, and that traditional coupon test standards can be misleading.

ACKNOWLEDGEMENTS

The authors would like to acknowledge the work and contributions of Karl-Fredrik Nilsson and Robin Olsson at FFA, and Mike Jones at Hurel-Dubois UK Ltd.

REFERENCES

1. Rouchon Jean, *Fatigue and damage tolerance aspects for composite aircraft structures*, Delegation Generale pour l'armement, Centre d'Essais Aeronautique de Toulouse, France
2. Olsson R, *DAMOCLES Task 1 - Deliverable: A Survey of Impact Conditions Relevant in Aircraft Composite Structures*, FFAP H-1353, FFA Report, 1998
3. Olsson, Beks, *Examination of impact response and damage of composite laminates*, FFA TN 1996-29, The Aeronautical Research Institute of Sweden, Bromma, 1996.
4. Morita, H. et al., *Characterization of impact damage resistance of CF/PEEK and CF/toughened epoxy laminates under low and high velocity impact tests*, Journal of Reinforced. Plastics and Composites, Vol.16, No.2, pp.131-143, 1997.
5. Olsson R, *Impact Response and Delamination of Composite Plates*, Doctoral Thesis Report 98-12, Royal Institute of Technology, Dept. of Aeronautics, Stockholm
6. G A O Davies & P Robinson, *Predicting failure by debonding/delamination*, Dept. of Aeronautics, Imperial College, AGARD: 74th structures and materials meeting, debonding and delamination of composites.
7. E Greenhalgh, S Singh & D Roberts, *Impact Damage Growth and Failure of Carbon-Fibre Reinforced Skin-Stringer Panels*, ICCM 11, Gold Coast, Brisbane, Australia, (1997)
8. E Greenhalgh, B Millson, R Thompson & P Sayers, *Testing and Failure Analysis of a CFRP Wingbox Containing a 150J Impact*, ICCM 12, Paris, France (1999)
9. Greenhalgh, E.S. and Singh, S., "*Investigation of the failure mechanisms for delamination growth from embedded defects*", Proc. 12th International Conference on Composite Materials (ICCM-12), Paris, France, 1999.
10. Nilsson, K., -F., Asp, L. and Alpman, J., "*Delamination buckling and growth at global buckling*", Proc. First International Conference on Damage and Failure of Interfaces, Ed. H.-P. Rossmanith, Vienna, 1997, pp193-202
11. Singh, S. and Greenhalgh, E.S., "*Micromechanisms of the interlaminar fracture in CFRP at multidirectional ply interfaces under static and fatigue loading*", Plastics, Rubber & Composites, Processing & Applications, Vol. 27, 1998.
12. Nilsson, K., -F. and Giannakopolus, A., "*A finite element analysis of configurational stability and finite growth of buckling driven delamination*", Journal of Mechanics and Physics of Solids, Vol. 43, 1995.

S695/003/99

S695/004/99

The hollow wide chord fan blade (WCFB) – a stiffened metallic structure

M McELHONE
Rolls-Royce, Derby, UK

ABSTRACT

Over the past 20–30 years, the thrust to weight ratio of aircraft engines has increased. At the same time, specific fuel consumption has been substantially reduced. A significant factor in promoting this trend has been the progression to larger fan diameters and bypass ratios. The fan blade of a high bypass ratio engine provides a challenging design task, requiring a compromise between the conflicting demands of aerodynamic efficiency, weight, vibration, noise, mechanical integrity and reliability, all for an acceptable production cost. Rolls-Royce has answered this challenge with the development of its unique hollow titanium wide chord fan blade (WCFB). The hollow wide chord fan blade is discussed, with particular reference to mechanical design considerations and method of manufacture.

INTRODUCTION

The aircraft gas turbine engine was developed, initially for military applications, in the 1940's, with the first civil applications following in the 1950's. During the Cold War period, military requirements drove the development of the technology and civil engines benefited from the spin-off and consequent reduction in development costs. Today's market, however, is very different. Military expenditure has declined and development cycles for new products, such as the EJ200 engine for the Eurofighter, have been significantly extended in time. The civil aero engine business is increasingly the focus of technological development as, in this highly competitive industry, the "big three" engine manufacturers, General Electric, Pratt and Whitney and Rolls-Royce battle for market share.

Continued growth in passenger traffic, together with the retirement of older aircraft, which are less environmentally acceptable and more costly to operate, generates a huge potential market for civil air transport products. However, competition in the airline business is fierce, with many airlines at best just profitable. Today's primary product discriminators are therefore cost driven, both initial procurement cost and cost of ownership. The challenge in the aero engine industry is to introduce technological developments, which improve engine performance while meeting these stringent cost targets.

THE MODERN CIVIL GAS TURBINE

The RB211 and Trent families of engines are central to Rolls-Royce's current civil product range, covering the range from 40,000 to 95,000 lb thrust. The Trent 800 engine (figure 1) is the largest member of the family to enter service with airlines. It weighs around 5 tons and passes more than one ton of air per second at the take-off condition, when one gallon of fuel is burned per second in the combustion chamber.

Engine thrust is related to mass flow and can be provided in two ways, either by moving a small mass of air quickly or by moving a larger mass of air more slowly. Military engines for fighter aircraft favour the former option, where agility and thrust-to-weight ratio are the important parameters. Civil engines, on the other hand, favour the latter, where efficiency and hence fuel consumption are critical. This improvement in efficiency is achieved using the bypass principle.

In a bypass engine, part of the air is compressed fully in the core engine and passes into the combustion chamber, where it is mixed with fuel and burned, before being expanded through the turbine. The remaining air is compressed only by the fan, before being ducted around the core of the engine to rejoin the hot gas downstream of the turbine. Here the hot gas-stream from the core and the cool gas-stream from the bypass duct mix together, resulting in a reduced overall jet velocity. Since propulsive efficiency is maximised when the jet velocity approaches that of the aircraft, this improves overall efficiency and hence reduces fuel consumption. An additional benefit of low jet velocity is a reduction in noise levels, an important factor in today's environmentally conscious world where noisy aircraft can be subject to restrictions when operating from certain airports.

The engine bypass ratio is the ratio of the bypass mass flow to core engine mass flow. Most modern civil gas turbine engines have high bypass ratios, a loosely defined term covering the range of bypass ratio from around 3 to 9. The bypass ratio of the Trent 800 engine is around 6. The choice of bypass ratio, as with all major engineering design decisions, is a compromise, in this case between performance and weight. A larger fan diameter will improve fuel efficiency, but at the cost of a reduced thrust-to-weight ratio. The choice of bypass ratio is also influenced by the aircraft application.

Each of the three major civil engine manufacturers faces the same technological challenges, which mainly arise from the requirements of wide-bodied aircraft. When these were first introduced, designs included mainly four and three-engined types such as the Boeing 747 and Lockheed L1011 Tristar. The modern trend, however, is towards a predominance of large twin-engined aircraft. The reasons behind this are economic, and can be illustrated by comparing the Airbus A330 and Boeing 777. The Boeing 777, with its two Trent 800 engines, can carry as many passengers as earlier versions of the four-engined Boeing 747. However, it achieves this at a much reduced cost due to the greater efficiency of its two larger engines over four smaller ones. The result of this trend to twin-engined aircraft has been an increase in the thrust requirement for each engine, which now exceeds 100,000 lb thrust. This is three times the thrust of the very first RB211 design and more than double that of the first engine for the Boeing 747.

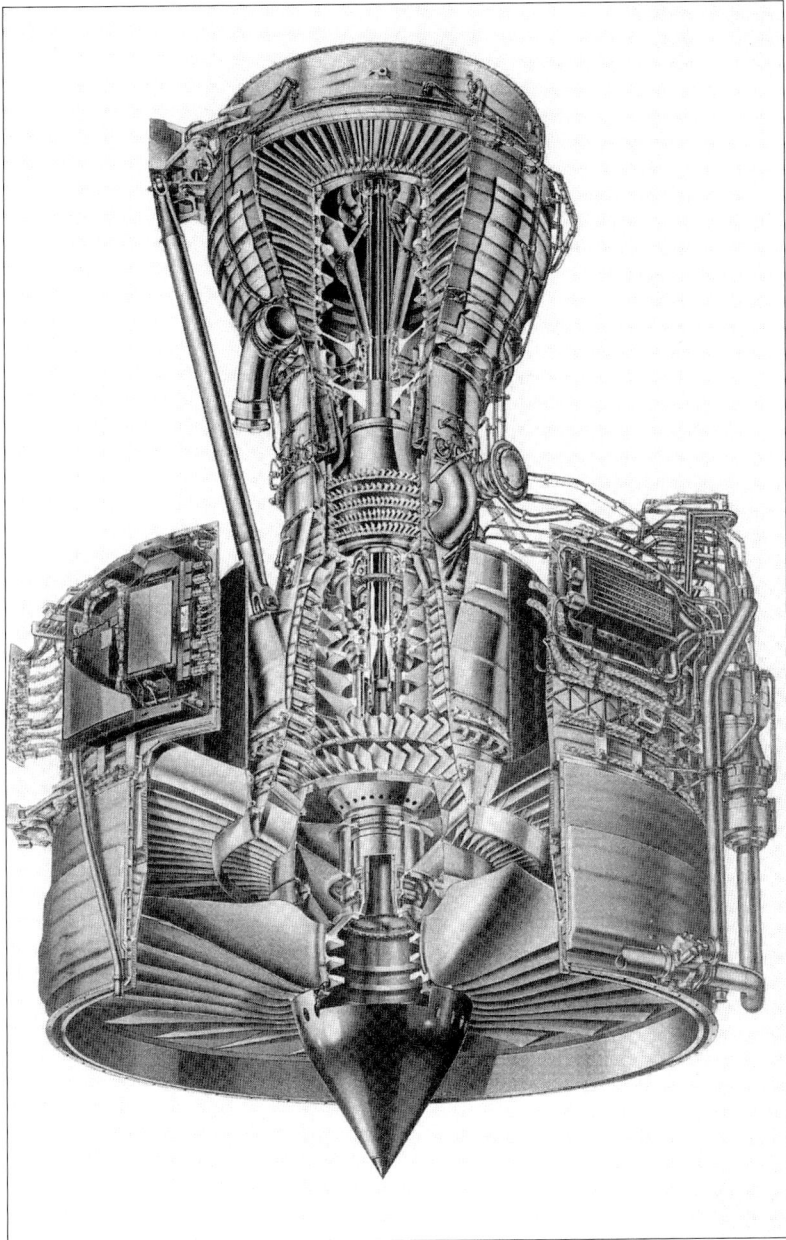

Fig 1 Rolls-Royce Trent 800

ENGINE DESIGN PARAMETERS

Take-off thrust is of paramount importance competitively. Several major airfields around the world suffer from restricted operations, which mean that the payload or range of the aircraft may be limited if take-off thrust is insufficient. This can be due to the fact that they have relatively short runways and obstacle in line with the flight path, for example London Gatwick and Hong Kong. Alternatively, since thrust reduces with air density, operation may be restricted because the airfields are at high altitude and generally high temperature, for example Johannesburg and Mexico City. To illustrate the importance of take-off thrust, consider the example of a Boeing 767 taking off from Hong Kong. A 1% increase in take-off thrust would permit an extra 4,500 lb aircraft weight to be lifted, which could mean additional revenue from around 14 extra passengers.

Once airborne, the aircraft needs to climb as quickly as possible to maximise the time spent at the more fuel efficient cruise conditions. At the top of climb, there needs to be sufficient thrust to achieve an adequate initial cruise altitude with a margin for operational flexibility. Rate of climb above 30,000 ft is a key design parameter, because the ability to achieve a high altitude at the start of cruise contributes to the fuel economy and removes the need for further periods of climb as fuel is used and the aircraft becomes lighter.

In a high bypass ratio engine, take-off thrust is provided mainly by the fan (figure 2). The key parameter, therefore, is fan mass airflow. Using two stages of blading in the fan would increase the mass flow, a strategy which is used in low bypass ratio military engines. However, this is impractical in a high bypass ratio civil engine, because of the interaction noise that would be produced by wakes in the airflow. Unfortunately, the pressure ratio of a single stage operating efficiently is limited. Airflow, and hence take-off thrust, is therefore determined by the fan diameter. Increasing physical size places considerable importance on design, for both low weight and structural stiffness.

The maximum fan diameter of the RB211 was 86 inches for many years and provided thrust developments from 33,000 to 63,000 lb, which is a remarkable achievement. There is a development cost advantage in maintaining a common physical size for as long as possible, and it is clear in retrospect that the absolute maximum was squeezed out in the evolution of this engine. The Trent 800 has a fan diameter of 110 inches, and fan aerodynamic improvements will eventually allow thrust developments to over 100,000 lb at this size. One of the Trent's direct competitors, on the other hand, has a much larger fan diameter of 123 inches and suffers a significant weight disadvantage as a consequence, despite being manufactured in carbon composite material.

S695/004/99 © IMechE 2000

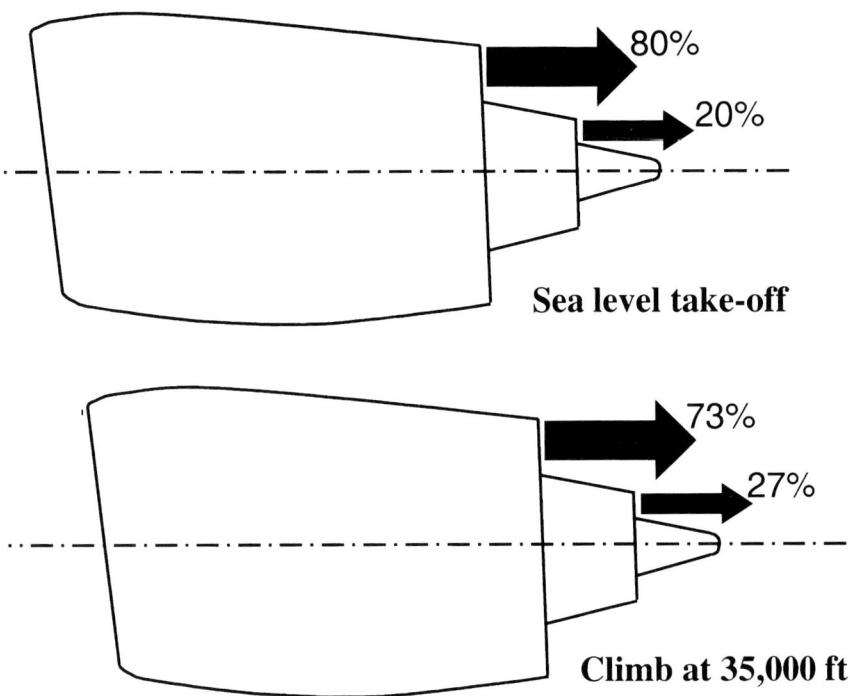

80%

20%

Sea level take-off

73%

27%

Climb at 35,000 ft

Fig. 2 Split of net thrust in a high bypass ratio engine

THE FAN BLADE

In the early days of gas turbine development, the major problems were largely with the combustor. In the case of compressor and turbine development, designers could draw on supercharger and steam turbine experience. However, with the combustor, there was little background experience on which to draw. As aero gas turbines grew in capability, the most difficult design and development task for many years was the high pressure turbine blade. This operates in an extremely harsh environment, which requires a careful compromise between the conflicting demands of aerodynamic efficiency, weight, vibration, and reliability. In addition, all this must be achieved at an acceptable and consistent production cost. In fact, turbine design is now highly advanced, with the high pressure turbine blade in a modern gas turbine engine operating in a gas stream whose temperature exceeds the melting point of the alloy used. It is only by the careful design of cooling passages and the use of surface coatings which prevents the blade from disintegrating.

Nowadays, the fan blade of a high bypass ratio engine provides an equally if not more challenging task. Since the fan blade can provide up to 80% of the engine thrust, optimisation of fan blade design is critical to achieving the primary requirement of maximised thrust and minimised fuel consumption. The same constraints of aerodynamic performance, weight, vibration and reliability as applied to the turbine blade, also apply to the fan blade. However, there are additional constraints of low noise and resistance to foreign object damage added to these. Satisfying the demands of mechanical integrity has become increasingly important as fan size has increased over the years.

The following numbers illustrate the scope of the design task. At take-off, the Trent 800 fan passes over 1.2 tons/second of airflow and compresses it through a pressure ratio of 1.8, with a circumferential speed at the blade tip of over 1,000 mph, which is 40 per cent faster than the speed of sound. The centrifugal (CF) load on each blade is around 90 tons, which is the same weight as ten London buses. Finally, the axial load on the fan assembly is over 20 tons and the power requirement of 89,000 shp is equivalent to around 1200 family cars.

Fan blades for high bypass ratio engines were, for many years, manufactured from solid titanium alloy forgings and were generally designed with mid-span snubbers (clappers) to control vibration. However, these snubbers impede the air flow, which is supersonic at this diameter, and therefore reduce the aerodynamic efficiency of the fan, leading to increased fuel consumption. By removing the blockage caused by the snubbers, fan efficiency can be increased by around 4%. However, removal of the snubbers can leave the blade susceptible to mechanical or aerodynamic instability. In the most modern designs, this problem has been overcome by increasing the blade chord to improve its vibration characteristics. The move to a wide chord design has reduced the number of blades in the assembly by around one third, which in conjunction with a unique hollow construction, incorporating an internal core, has resulted in a significant weight reduction (figure 3).

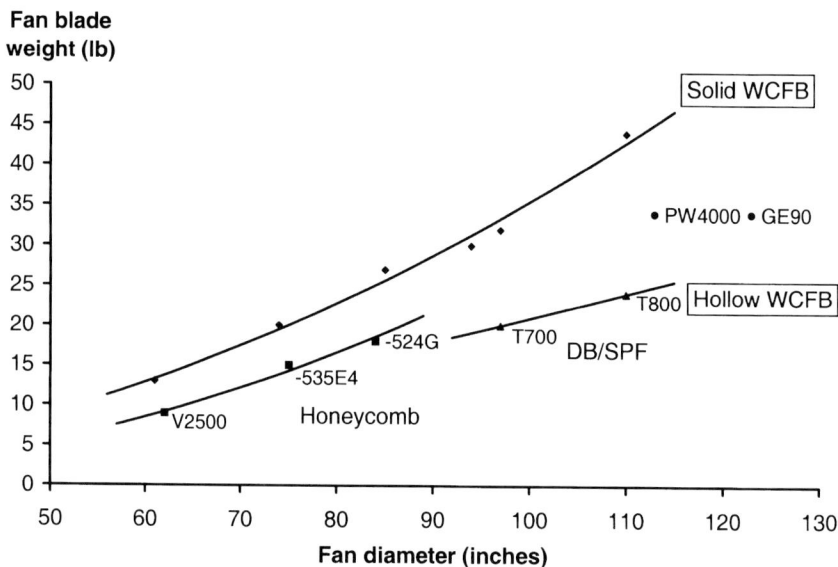

Fig. 3 Fan blade weight against fan diameter for solid and hollow blades

S695/004/99 © IMechE 2000

THE "FIRST GENERATION" HOLLOW WIDE CHORD FAN BLADE

The preparatory work which had to be completed satisfactorily before this design could be committed to production covered about 8 years and is an excellent example of simultaneous engineering i.e. design for a controllable manufacturing process as well as for the product duty. It was started shortly after the original carbon-composite wide chord design for the RB211 was replaced by a snubbered solid titanium blade.

The low-density core for the hollow design is an integral part of the structure. The basic construction of the blade is that of an activated diffusion bonded titanium shell with a lightweight honeycomb core. It is essential that both the panel-panel and core-panel joints achieve parent material properties, to enable them to withstand the effects of foreign object damage and fatigue.

The manufacturing process begins with two titanium panels, with stub ends for the root block, which are twisted and then hot creep formed in dies which closely match the finished blade form. Chemical milling operations then progressively hollow out the inside of the panels to receive the titanium honeycomb core, which is pre-machined to the shape of the resulting cavity between the panels. The three component parts are then assembled and heated in an evacuated shell, where the joints are formed by a transient liquid phase diffusion bonding process. After hot resetting between dies, to correct any distortion introduced during the bonding process, the leading and trailing edges, tip and root are then machined to shape, prior to final surface treatment.

THE "SECOND GENERATION" DB/SPF BLADE

Whilst the first generation blade led the field in technology of the day, and has given an excellent service record, it is complex and labour intensive to produce. In addition, the stress concentration feature at the honeycomb to panel interface has to be allowed for in the design, and the honeycomb core cannot support its own weight, necessitating a thicker panel design with an associated weight penalty.

The basic concept of the second generation blade was to replace the core with a more self supporting structure, and to avoid the complexity of an activated diffusion bond. This would result in a lighter and cheaper blade, with reduced manufacturing lead-time. The underlying philosophy of the second generation blade was still that of a titanium shell with a lightweight core. However, the second generation blade employs solid-state diffusion bonding (DB) in association with superplastic forming (SPF) of the assembly to produce a corrugated internal core structure (figure 4).

Fig. 4 Wide chord fan blade core construction

It is an elegant process (figure 5), whose simplicity is sometimes not fully appreciated by the non-engineer. An inhibiting compound is applied to the mating faces of the external panels in a pattern derived from the design of the corrugation, so that diffusion bonding can occur only at surfaces not coated with inhibitor. The three piece "flat-pack" sandwich is then bonded in a custom built high temperature pressure vessel, which allows joining at all surfaces not coated with the inhibitor. Tight process control guarantees the generation of diffusion bonds with an appropriate microstructure and therefore acceptable mechanical properties.

After bonding, the pack is then converted to an aerofoil shape via a sequence of hot forming operations. The manufacturing sequence then exploits the superplasticity of fine grain titanium alloys. The cavity of the bonded construction is inflated at elevated temperatures between contoured metal dies, using an inert gas to expand the core and simultaneously develop the blade's external aerodynamic profile. Machining of the external surfaces and final surface processing then completes the manufacturing process.

Considerable experimentation was necessary to establish the process parameters required to produce blades of consistent shape and properties and to satisfy the rigorous quality standards required for the service environment. All blades are inspected with a combination of ultrasonic, radiographic and more conventional methods. These inspections have, of course, enabled the fabrication processes to be fully understood and matched with the requirements of the design specification.

MECHANICAL INTEGRITY

The reliability and integrity of the hollow wide chord fan blade, which was first introduced on the RB211-535E4 engine almost 15 years ago, has been second to none. The step in technology at its introduction produced a major competitive advantage and 10 years passed before an equivalent design appeared from another manufacturer. This is an unusually long period for such a step to remain unmatched in the highly competitive aero-engine industry.

                                                  S695/004/99 © IMechE 2000

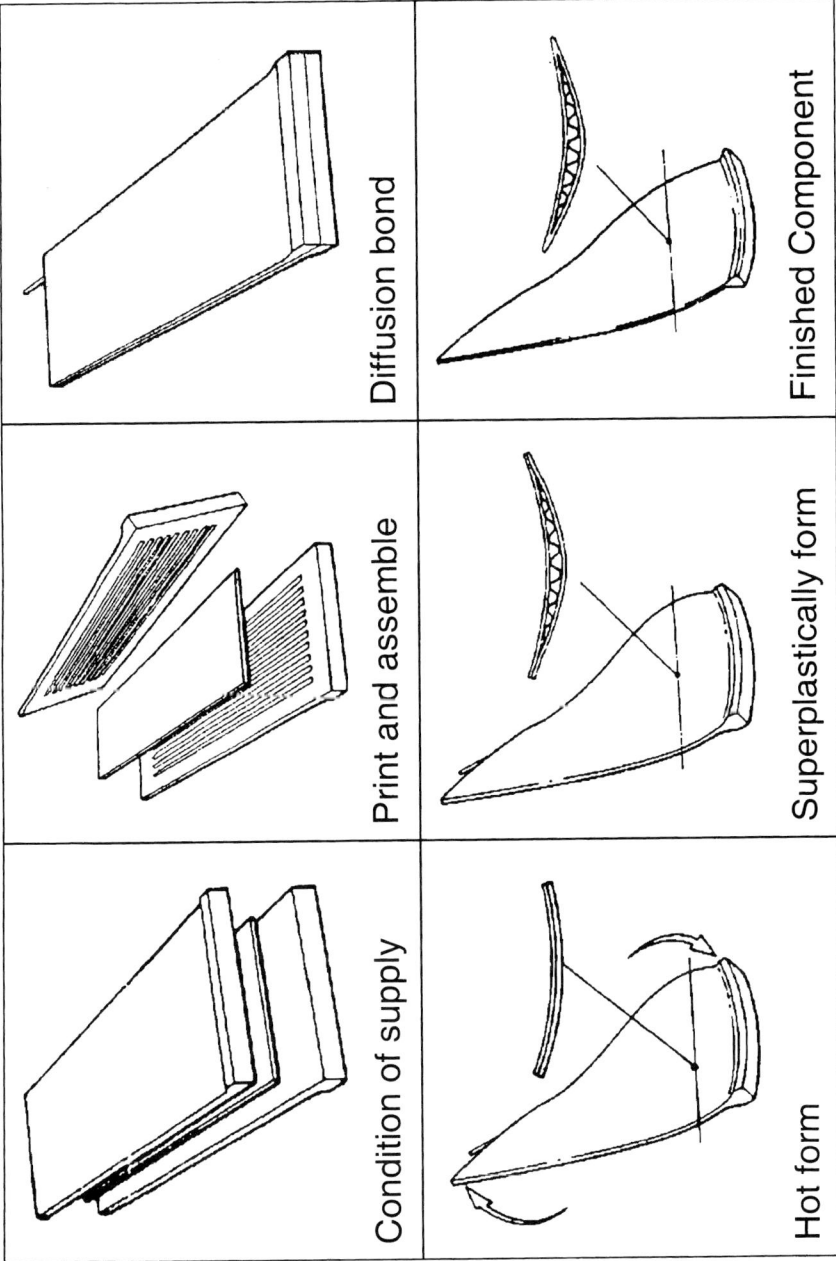

Condition of supply

Print and assemble

Diffusion bond

Hot form

Superplastically form

Finished Component

Fig. 5 DB/SPF fan blade manufacturing sequence

The excellent service record is the result of a combination of careful finite element (FE) analysis, together with the most thorough development testing and airworthiness demonstrations. The company standard FE analysis package is SC03, which was developed in-house and is used for blade and disc stress and vibration analyses (figure 6). During the blade design phase, stress and vibration analyses are carried out using SC03, and compared with allowable design criteria to ensure that the project life and integrity requirements are achieved.

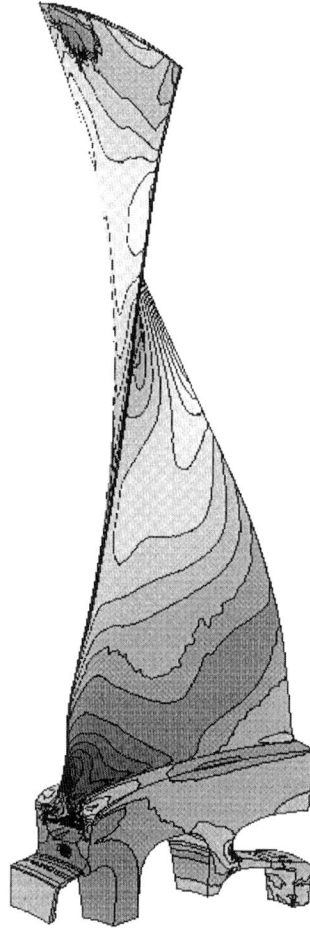

Fig. 6 SC03 blade and disc stress analysis

In addition to detailed stress and vibration analysis, fatigue testing in both low and high cycle modes is carried out. Groups of blades are repeatedly accelerated to maximum speed in vacuum to establish their low cycle fatigue endurance. Similarly, high cycle fatigue endurance is investigated on a static vibration rig, to stress levels which exceed the most severe service conditions encountered, which generally occur during cross-wind operation.

With such a large forward facing area, foreign object ingestion must be considered seriously, and resistance to bird ingestion is a particularly demanding requirement. Medium sized birds tend to congregate around airfields, particularly in coastal areas, whereas larger birds can be encountered during cruise, flying singly up to remarkably high altitudes. Ingestion of a number of medium sized birds during take-off usually represents the more severe design requirement.

Blade resistance to birdstrike can be modelled using the commercially available Dyna3D package, running on powerful graphical workstations linked to a supercomputer. It is a requirement of the airworthiness authorities, however, that birdstrike resistance is adequately demonstrated during the engine development programme. The ability to survive multiple bird ingestion had to be demonstrated in the case of the Trent 800 by running the engine up to take-off power and requiring it to ingest four birds of 2 ½ lb weight within one second. The kinetic energy involved is equivalent to a ten-pin bowling ball travelling at around 80 mph. Following the bird ingestion, the engine continued to deliver power, accelerate and decelerate for a total period of 30 minutes, to simulate the likely operating procedure following a severe ingestion incident. Thirty minutes would be sufficient to allow the aircraft to climb to a safe altitude, circle for a time to off-load fuel and then return to make a safe landing.

In the very unlikely event of a blade mechanical failure, perhaps initiated by damage from the ingestion of a hard object, the engine has to be shown to be structurally sound and to contain all the debris, even if the failure occurs at maximum power. Fan blade containment in modern engines is achieved with aluminium or titanium casings through which the blades can penetrate, to be then caught in external windings of Kevlar fabric. The sequential elements of blade failure and the subsequent behaviour of the engine require the most careful design consideration. These include the protection of adjacent blades, the collection of the debris in the Kevlar windings, the deflection of the Kevlar as it absorbs the energy loads, the integrity of the structure under the highly unbalanced conditions which result, and finally the integrity of the externally mounted accessories which must not be allowed to release fuel, oil or hydraulic fluid and cause the danger of fire. Again, Dyna3D is used to simulate these complex dynamic events and again, as with bird ingestion, the airworthiness authorities require that engine integrity is ultimately demonstrated on a development engine.

The product of this design and development effort has been a significant contributor to the Trent 800's competitiveness. The hollow wide chord fan blade has a very high flow for its frontal area, as a consequence of its high aerodynamic efficiency. In addition, due to the lightweight corrugated core, the Trent 800 DB/SPF fan blade provides a significant contribution to the overall weight saving on this engine. The weight saving due to the DB/SPF blade is around 15% for the blade alone. When other whole engine effects are taken into account, the Trent ship set is some 7,200 lb and 5,300 lb lighter than the competition (GE90 and PW4090 respectively). This equates to around a 1% improvement in specific fuel consumption or, alternatively, increased payload capacity.

FUTURE DEVELOPMENTS

Looking to the future, some, for instance General Electric, believe that carbon composite materials can be used to reduce weight. At present, these materials limit the speeds for which the blade can be designed, requiring therefore a greater fan diameter for a given thrust. It may be that this alternative approach, as it has been developed so far, will converge in principle with the hollow design because airworthiness requirements have led to the incorporation of titanium sheathing around a proportion of the composite blade, with an associated weight penalty. The composite could therefore be considered as an alternative core to the titanium honeycomb or corrugation.

For the foreseeable future, Rolls-Royce plans to continue with its well proven hollow titanium concept and is confident that it will continue to be competitive on weight. However, the possibility of using titanium metal matrix composites is being considered. These would allow selective reinforcement of the fan blade with silicon carbide fibres, to modify blade untwist and vibration characteristics.

Aerodynamic development continues aggressively, aiming at higher specific flow to reduce engine frontal area and weight, and at improving aerodynamic efficiency over the wide range of operating conditions that have to be met. The use of 3-dimensional computational fluid dynamics (CFD) methods has led to the design of a swept blade, similar in principle to the sweep of an aircraft wing (figure 7). The Trent 8104 project is committed to a swept fan design for this engine, which recently broke the world record by exceeding 110,000 lb thrust during development testing. Work in this area will continue, in order to maintain our competitive edge and ensure that we achieve satisfactory mechanical condition and integrity.

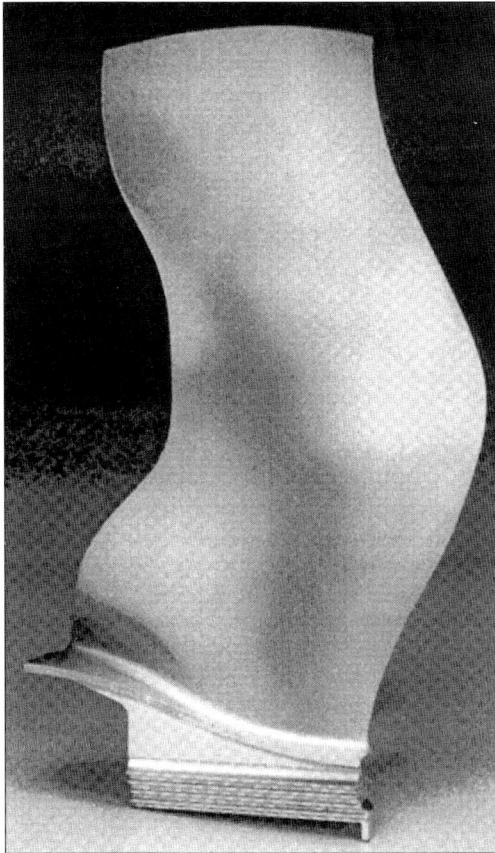

Fig. 7 Swept fan blade prototype

CONCLUSIONS

The fan blade is now one of the main focuses of competitive attention in high bypass ratio technology. Widely differing concepts are evident within the industry, but the same constraints on aerodynamic efficiency, weight, noise, vibration, reliability and integrity apply to all the manufacturers. Rolls-Royce has answered this challenge with the hollow titanium wide chord fan blade, which is based on over 20 years of technological advancement and progressive development. The hollow wide chord fan blade stands up well to any comparison, and current and future developments will ensure that Rolls-Royce remains at the forefront of the highly competitive aero engine industry.

S695/005/99

Unidirectional carbon-fibre rods for high-performance structures

M R WISNOM and **K D POTTER**
Department of Aerospace Engineering, University of Bristol, UK

ABSTRACT

The characteristics of Graphlite precured rods and ways in which the rods can be incorporated in structures are discussed. The concept of overwinding rod elements with Kevlar under very high tension is presented. It is shown that this suppresses splitting and sensitivity to impact, leading to a highly damage tolerant form of material.

INTRODUCTION

Carbon fibre composites have very high specific strength and stiffness in the fibre direction. Recent research at Bristol University has been looking at innovative approaches to composite structures in order to take greater advantage of these properties. One aspect of this work has been to investigate the mechanical properties of Graphlite rods. Test results are presented here that show extremely high tensile and compressive strengths for single rods. Ways of using the rods by incorporating them into plates, tubes, stiffeners and highly anisotropic components are discussed. A second aspect of the research has been the investigation of a novel approach to improving damage tolerance by overwinding rod elements with Kevlar under very high tension. Results are presented for impact tests on overwound and bare rods at different levels of incident energy which show that the overwinding eliminates splitting. After impact a dent is clearly visible on the surface of the Kevlar, but there is little reduction in residual compressive strength.

GRAPHLITE RODS

Graphlite is a precured unidirectional carbon fibre-epoxy rod produced in various sizes with different fibre and matrix systems. It is manufactured by a modified pultrusion process that produces excellent fibre alignment, resulting in high compressive strength. A key advantage of the material is that the alignment can be maintained when the rods are incorporated into structures because the rods are precured. With conventional pre-preg materials wrinkling and kinks often develop during the manufacturing process, which can lead to large reductions in strength.

The properties of single Graphlite rods have been measured by Clarke using specially developed test methods [1]. Rods of 0.7 mm diameter made from IM7 intermediate modulus carbon fibres were tested in tension and compression. The volume fraction was 63.2%. The geometry of the tensile test specimens is shown in Fig. 1. The end tabs were produced from aluminium bar, with the end 5 mm drilled accurately to the same diameter as the rods to maintain the rod centrally. The remainder of the length was drilled out to a larger diameter to allow a uniform thickness of adhesive.

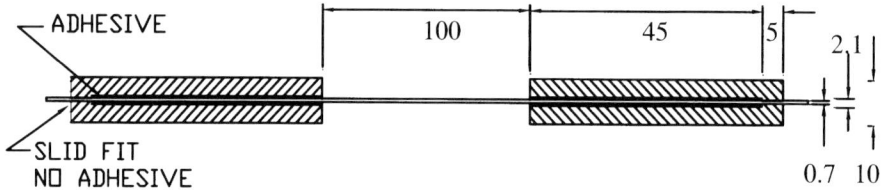

Fig. 1: Tensile test specimen for Graphlite rods (dimensions in mm)

A relatively thick bondline (0.7 mm) and a ductile, high strain adhesive (3M 9323 A/B) were used to try to minimise stress concentrations. The specimens were laid in a V block to maintain alignment during curing.

Specimens were carefully aligned and gripped in the jaws of an Instron servohydraulic test machine, and tested to failure. The average strength of 10 specimens was 3.70 GPa. Failure was sudden, destroying most of the specimen. It appeared to initiate near the end tabs, and may therefore underestimate the true ultimate strength of the material.

Rods were tested in compression using the cylindrical end tabs shown in Fig. 2a. This time the closely drilled sections were placed at the ends of the gauge section and were coated with release agent to effectively provide anti-buckling guides to the specimen. The tabs were made of silver steel, and the rods were cut short to prevent direct compressive loading on the ends. Specimens were tested in a compression fixture consisting of a linear die set to maintain alignment whilst minimising friction, as shown in Fig. 2b. The adhesive was cured with the specimen mounted in the die set to avoid misalignment or stresses being imposed during set-up.

S695/005/99 © IMechE 2000

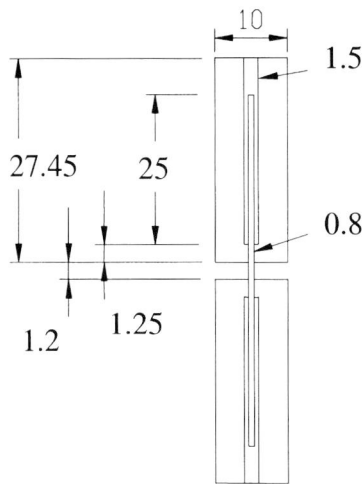

a) Specimen

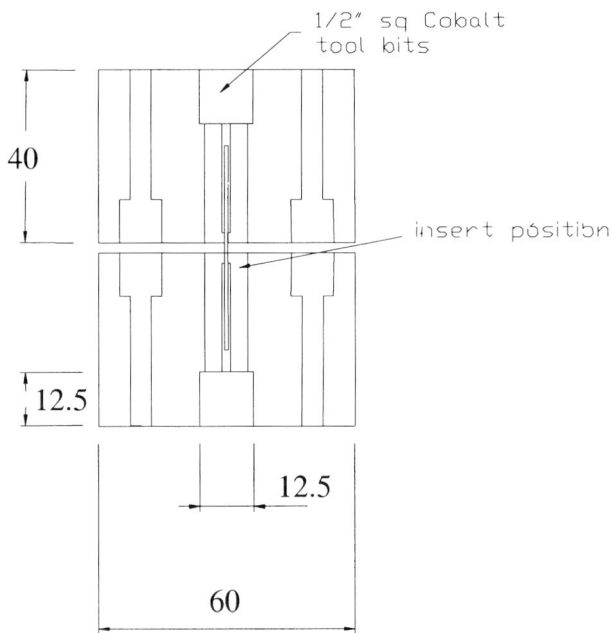

b) Test fixture

Fig. 2: Compressive test arrangement for Graphlite rods (dimensions in mm)

The average compressive strength from 10 specimens was 2.43 GPa. A small amount of sliding was observed as the adhesive yielded, and then failure occurred suddenly, with a kink band in the gauge section.

Both the tensile and compressive strengths are very high compared with typically measured values for unidirectional carbon fibre-epoxy, showing the excellent performance that can be obtained with the Graphlite rods. One limitation with the rods tested here is the relatively low glass transition temperature of about 110°C. However, use of higher temperature resins is under investigation.

INCORPORATION OF RODS IN STRUCTURES

Plates have been manufactured at Bristol University by incorporating Graphlite rods between layers of pre-preg [2]. Two plies of unidirectional T800/924 carbon fibre-epoxy were placed on either side of a single row of rods, as shown schematically in Fig. 3. T800 is also an intermediate modulus fibre with a similar modulus to the IM7 fibres used in the rods. Good consolidation was achieved, with the pre-preg filling the spaces between rods with minimal voidage. The process produced a plate 1.1 mm thick, with an average fibre volume fraction of 62%.

Specimens 10 mm wide were cut and bonded to 2 mm thick angle ply glass-epoxy tabs. Reverse tapers were machined on the tabs, and large adhesive fillets were used to try to minimise the effect of stress concentrations. Tension tests were carried out by gripping specimens with 100 mm gauge length directly in the test machine. Compression tests used a gauge length of 5 mm to avoid buckling, and were performed in the Imperial College test fixture [3].

Results were disappointing, with mean tensile and compressive strengths of 2.38 GPa and 1.46 GPa respectively, only 64% and 60% of the values achieved in the single rod tests. The corresponding strains at failure measured with strain gauges were 1.38% and 1.11%. These values equate to secant moduli of 172 GPa and 132 GPa in tension and compression. Some non-linearity was observed in the stress-strain response, with stiffening in tension, and softening in compression, hence the difference in secant moduli at failure.

It is believed that the lower strengths were primarily due to difficulties with introducing the load in the tests. There was also a suspicion that the outer layer of pre-preg may have contributed to premature failure. These concerns are supported by the much higher strains achieved in tube tests reported later. However, some of the difference could be due to size effects, since the volume of material in the single rod tests was very small. Further research is needed to investigate this.

An alternative method of incorporating the rods into plates was also investigated using only one layer of pre-preg on the outside, and replacing the inner layers by FM300 film adhesive. This produced plates of similar quality with a thickness of 1.035 mm. The fibre volume fraction was much reduced at 42.8%. Strains at failure in tension and compression were 1.47% and 1.08%, similar to those for the plates not using film adhesive. Strengths were much lower at 1.78 GPa and 1.00 GPa due to the lower volume fraction.

S695/005/99 © IMechE 2000

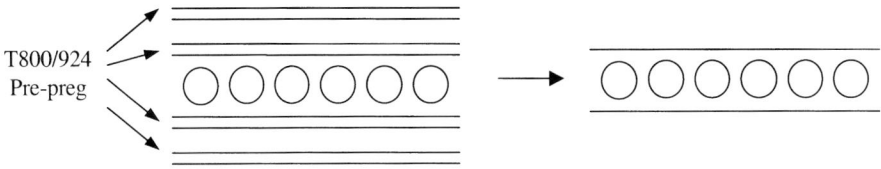

Fig. 3: Schematic of layup of Graphlite plates

Tubes have also been made using a similar layup process [2,4]. A single layer of pre-preg was rolled onto a mandrel, followed by two layers of rods sandwiched between three layers of film adhesive. A final layer of pre-preg was placed on the outside, using E glass/913 epoxy instead of T800/924 to avoid possible problems due to failure initiating in the surface carbon fibre layer. The final internal and external diameters were 9.9 and 13.5 mm respectively.

Tubes 680 mm long were tested in a pin-ended buckling rig to produce flexural failure whilst avoiding problems due to stress concentrations at the loading points [5]. Initial results showed some problems due to ovalisation of the cross-section under the very large displacements that were reached. In later tests the tubes were filled with polyester resin to suppress this unwanted behaviour. Strain gauges were placed near the centre of the tube at positions furthest from the neutral axis to record the maximum strains. Fig. 4 shows the test set-up schematically.

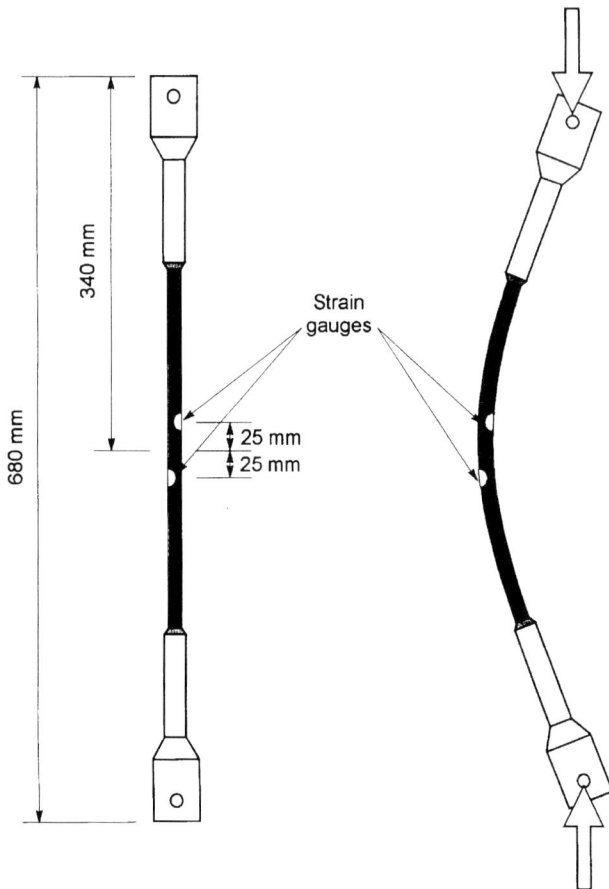

Fig. 4: Pin-ended buckling test for Graphlite tubes

The tubes failed suddenly in compression near the centre, with strains as high as 1.79%. This equates to a compressive stress of 2.4 GPa in the rods based on the previous tests on plates incorporating Graphlite rods, assuming linear elasticity and direct proportionality between volume fraction and modulus. This is very similar to the strength achieved in the single rod tests, although the true stress in the tubes would in fact be slightly lower due to the non-linear stress-strain response. High tensile strains of over 1.4% were also withstood without failure. Tubes with two layers of +/-45° glass-epoxy on the outside instead of the single layer of unidirectional material were found to behave similarly. These results show that the excellent properties of the rods can be translated into very high performance in components made from them.

S695/005/99 © IMechE 2000

Bell Helicopters have used Graphlite in stiffened skin panels for the V-22 tilt rotor. Fig. 5 shows a picture of a stringer incorporating the rods in the top and bottom flanges [6]. The rods were embedded in Syncore, a syntactic epoxy resin filled with hollow ceramic micro-spheres, and +/-45° carbon-epoxy pre-preg plies were used on either side of the layers of rods to transfer the load into them. Bending tests on sections of stringer where the small rod pack was loaded in compression produced strains of about 1.2%. This form of construction was also found to be damage tolerant, able to sustain compressive strains above 0.6% after impact damage.

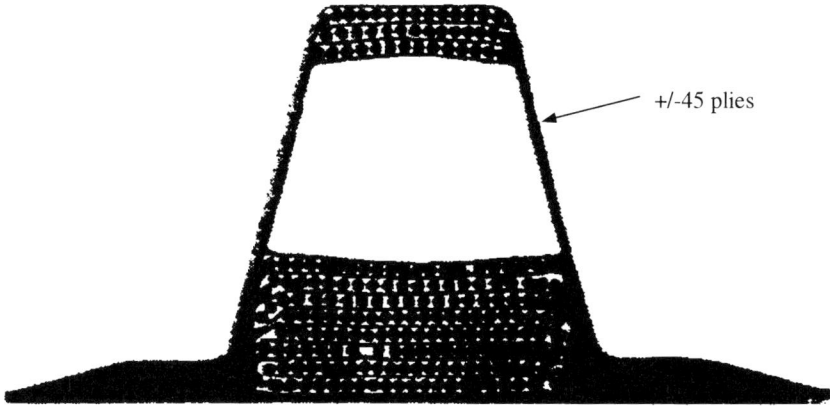

+/-45 plies

Fig. 5: Cross-section of stringer incorporating Graphlite rods [6]

Potential cost savings have also been identified based on the ability to place large quantities of unidirectional material accurately and in a short time with mechanised material handling. The rods lend themselves to layup on lightly curved surfaces. Since they are fully cured there is no shelf life limitation on material storage.

The properties of Graphlite also open up other possibilities for using them in novel structural configurations. One example of this is in producing highly anisotropic materials by embedding the rods in a flexible matrix. Aligning all the rods in one direction can produce high axial and bending stiffness and strength combined with low shear and torsional stiffness. Alternatively by laying rods at +/-45° high shear or torsional stiffness and strength can be achieved with low axial and bending stiffness.

The first approach has been investigated at Bristol University. Rectangular specimens made up of arrays of Graphlite rods in a matrix of epoxy blended with polyurethane have been manufactured using resin transfer moulding. The rods formed 60% of the volume of the specimens. Tests gave a flexural modulus of 82 GPa and showed that strains of 1.1% can be achieved without any local buckling of the rods whilst producing an effective shear modulus in torsion of only 0.83 GPa.

A larger scale demonstration component was produced for a helicopter application where a specified low value of torsional stiffness was required. The component was bowtie shaped, as shown in Fig. 6a. It was 800 mm long, with a rectangular cross-section of 104 x 46 mm at the ends tapering down to 50 x 22mm at the centre. End fittings were produced from aluminium alloy blocks carefully drilled with individual holes to accept the 1.7 mm diameter rods. Aramid fibres were woven through the rods at two points along the length to maintain the correct cross-section. The rods were bonded into the end blocks, the component was placed into a mould tool and resin was injected.

The component had a very high bending stiffness, and met the low torsional stiffness requirement. Tests were carried out at up to 20° angle of twist, with no damage. Fig. 6b shows the component under load.

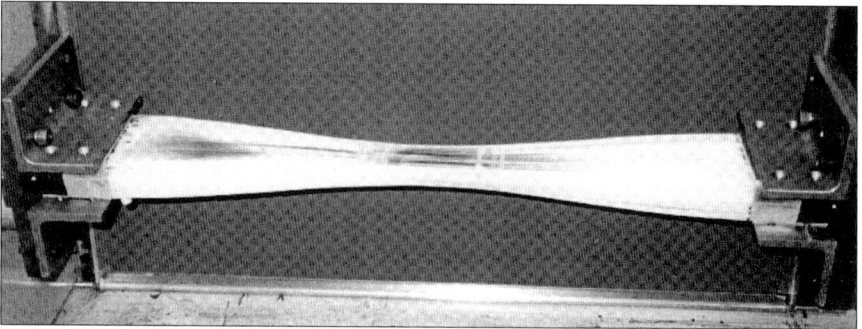

Fig. 6a: Graphlite bowtie specimen mounted in the torsion machine

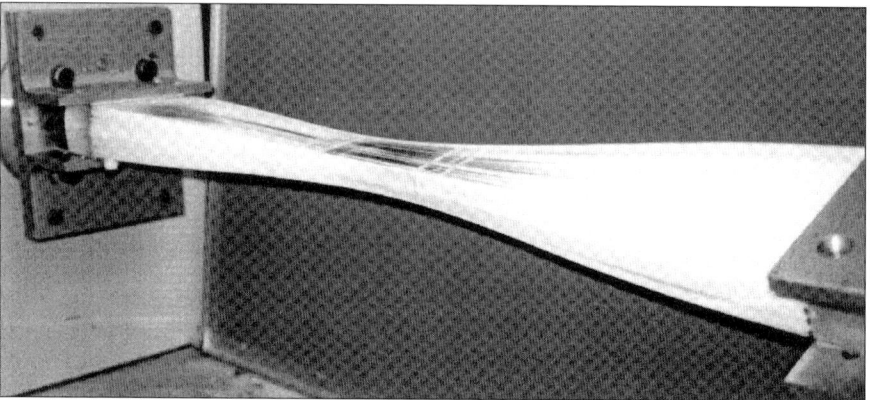

Fig. 6b: Graphlite bowtie specimen with 20° of twist

    S695/005/99 © IMechE 2000

OVERWINDING RODS WITH KEVLAR UNDER TENSION

The concept of overwinding unidirectional carbon fibre with high strength aramid fibre has been proposed as a way of overcoming the susceptibility of the material to splitting and low compressive strength after impact [7]. Due to its excellent tensile strength, the aramid fibre can be wound under very high tension, inducing transverse compression in the carbon which effectively adds to the tranverse strength. Even if there is damage, the presence of the overwinding inhibits propagation. The results of compression tests after different levels of impact on rods with and without overwinding are presented in order to demonstrate what can be achieved with this approach.

Suitable large diameter rods were not available, and so solid rods of 12.4 mm diameter were manufactured from 240 0.7 mm diameter Graphlite rods set in MY750 epoxy. This gave an overall volume fraction of carbon of about 47%, and a modulus of about 123 GPa. Seven layers of Kevlar 49 tows were wound on at an angle of 87 degrees, producing a final winding thickness of 1.55 mm. The winding was done under tension using a system of pulleys and a dead weight giving a stress in the Kevlar of about 450 MPa.

Sections of overwound and bare rod were clamped in fixtures which effectively built in both ends, leaving a free length of 50 mm. They were impacted by dropping steel weights with a 10 mm diameter cylindrical surface from various heights to give incident energies of up to 60 J. Compressive tests were then performed to determine the residual strength. The Kevlar was removed from some of the overwound rods to determine the state of damage after impact.

The bare rods split in two at an impact energy level of 10 J, and suffered extensive damage with multiple splits at 40 J. At this level of impact the residual compressive strength was only 32% of the unimpacted value. The overwound rods showed no splitting, even at an energy level of 60 J. At 40 J there were signs of damage in the form of partial compressive creases on the surface on each side of the impact zone. In all cases there were clearly visible dents in the Kevlar at the point of impact.

The compressive strength of the overwound rod with no impact was significantly higher than for the bare rod. This may be because the overwinding reduced the discontinuity in lateral support as the rod emerges from the very stiff end fixtures. The residual strength after impact reduced slightly with increasing energy, but even at 40 J was greater than that of the unimpacted bare rod.

These results are summarised in Fig. 7. Failure stresses are based on the cross-sectional area of the carbon fibre rod ignoring the Kevlar, and normalised to 60% volume fraction. The overwound rods continued to carry large compressive stresses after the point of maximum load, in marked contrast to the sudden, catastrophic failure of the bare rods. Post failure inspection revealed that kink bands had propagated stably in a zig-zag pattern along the gauge length, contained by the Kevlar overwind.

1200
1000
800
600
400
200
0

Normalised compressive strength. MPa

Overwound samples

Unwound samples

0   10   20   30   40

Impact energy J

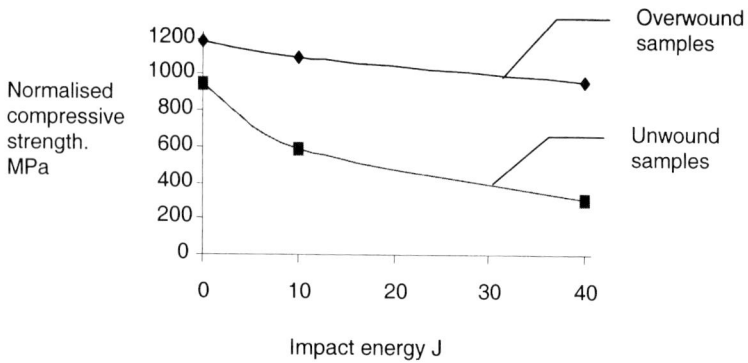

Fig. 7: Effect of overwind on impact performance

The radial compressive stress in the rod after winding was estimated to be about 36 MPa based on a strain gauge in the hoop direction on the surface of one rod. Lower winding tensions with the same amount of Kevlar gave greatly reduced strengths, confirming that it is the stress induced by the overwinding that is primarily responsible for the improved performance.

The overwinding completely suppressed splitting under impact loading, and allowed the material to reach its fibre direction compressive failure stress without transverse failure occurring. There was clear evidence of impact in the form of a dent in the overwound fibres, although it had little effect on residual compressive strength. This is a very desirable characteristic, and is the opposite of the normal situation with carbon fibre composites where impact damage that may not be visible can have a severe adverse effect on strength.

Excellent performance was achieved with a radial compressive stress of about 36 MPa. The increase in mass was 43%. The winding tension was restricted by limitations of the available equipment, and with some further developments it should be feasible to increase it to at least 1000 MPa. With the use of an increased volume fraction for the carbon fibre, it may be possible to produce an overwound rod that is actually lighter than an equivalent unwound rod. Calculations indicate that a 70% volume fraction rod overwound at 1000 MPa to produce compression of 36 MPa would be about 5% lighter than an unwound rod of 60% volume fraction with the same axial stiffness and nominal load carrying capacity [7].

CONCLUSIONS

The tensile and compressive strengths of 0.7 mm diameter Graphlite rods made from intermediate modulus carbon fibre-epoxy were measured to be 3.7 GPa and 2.4 GPa, both very high compared with typical values for carbon fibre composites. The high compressive strength is believed to be due to the excellent fibre alignment, which should be maintained when the rods are incorporated into structures because the rods are precured.

S695/005/99 © IMechE 2000

Rods can be laid up with pre-preg to produce plates, tubes and stiffeners. Strains of 1.79% measured on tubes show that the excellent properties of the rods can be translated into very high performance in components made from them. There is considerable scope to exploit the properties of the rods in novel ways, such as in composites of extreme anisotropy made by embedding rods in a flexible matrix.

Overwinding rod elements with high strength aramid fibre under tension suppresses splitting and impact sensitivity. Under impact loading the material is able to achieve its full compressive strength before transverse failure occurs. Residual compressive strength is greatly improved, and shows only a small reduction with increasing impact energy. The failure mode is not catastrophic as the overwind is able to contain the propagating kink bands. Impact damage can be seen on the surface of the overwound fibres, but has little effect, eliminating the problem of barely visible impact damage.

ACKNOWLEDGEMENTS

This work was supported by the U.K. Engineering and Physical Sciences Research Council under contract GR/K66833 in collaboration with British Aerospace, GKN Westland Helicopters, Cookson, Reaction Engines and Cranfield University.

REFERENCES

[1]   Clarke, A., Wisnom, M. R. and Potter, K., A comparison of the mechanical properties of Graphlite unidirectional carbon rod with conventional prepreg, Plastics, Rubber and Composites Processing and Applications 26: 447-450, 1997.

[2]   Clarke, A. B., Mechanical properties and process conversion of a novel form of unidirectional carbon fibre/epoxy rod, PhD thesis, University of Bristol Department of Aerospace Engineering, November 1998.

[3]   Haberle, J. D. and Matthews, F. L., An improved technique for compression testing of unidirectional fibre-reinforced plastic; development and results, Composites 25:358-371, 1994.

[4]   Clarke, A., Wisnom, M. R., and Potter, K., Development of high strength unidirectional carbon tubular elements from small diameter precured rod, Proc. FRC98, Newcastle, April 1998, pp. 45-52.

[5]   Vaughn, L. F., Potter, K., Clarke, A. B. and Wisnom, M. R., Compressive testing of high performance carbon composite tubes, ECCM8, Naples, Vol. 1, pp 315-322, June 1998.

[6]   Rogers, C. and Raczelowski, S., Design Guidelines for Rod Reinforced Structure, NEPTCO Inc., March 1995.

[7]   Wisnom, M. R., Suppression of splitting and impact sensitivity of unidirectional carbon-fibre composite rods using tensioned overwind, Composites Part A, 30:661-665, 1999.

S695/006/99

Stiffened composite structures for aviation and marine applications

N G FOSTER
Slingsby Aviation Limited, Kirkbymoorside, UK

1   INTRODUCTION

Slingsby Aviation specialises in the manufacture of composite aviation and marine structures. These structures, are optimised for strength, stiffness and weight and therefore skin thicknesses are usually low. In order not to increase aerodynamic and hydrodynamic drag loads, the structural surfaces need to be stable up to limit load and sometimes up to ultimate load.

This paper presents some of the methods used to stabilise thin skinned aviation and marine composite structures.

The types of structure described in this paper are:

i)   Sandwich structures consisting of thin composite skins stabilised with a closed cell foam core.

ii)   Sandwich structures consisting of thin composite skins stabilised with an open hexagonal aramid fibre core (Nomex).

iii)   Composite skins stiffened internally with composite top hat stiffeners, 'C' channel ring frames and corrugations.

iv)   Anechoic cores sandwiched between composite skins.

Some of the composite structures described in this paper replace structures manufactured in aluminium, steel and wood.

2   THE T67 FIREFLY

The T67 Firefly (Fig 1) is a two seat composite aircraft, developed from the Fournier RF6B which was manufactured from wood. When the aircraft was redesigned in glass fibre, the thin skins of the fuselage, wing, tailplane and control surfaces were stiffened and stabilised using conventional methods. The aircraft is fully aerobatic and is stressed to +6g and -3g with a life of 18,000 hours. The fuselage and wing have been statically tested at 60°C to the equivalent of 11g following a fatigue test at ambient temperature. There are a number of variants powered by the 160, 200 and 260 Hp Lycoming engines. The Firefly is operating with the RAF, the USAF, the Canadian Aviation Training Centre, the KLM Flying Academy of the Netherlands and the Turkish Air League.

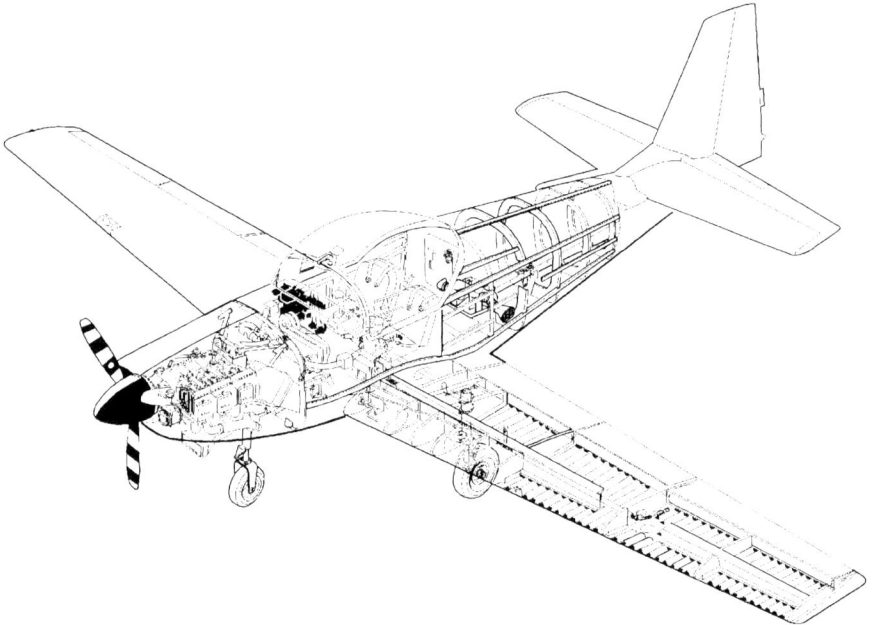

Figure 1 T67 firefly

The fuselage, which is effectively a closed box section with a cut-out for wing and canopy, carries bending, shear and torsion loads. The upper and lower longerons carry the fuselage vertical and lateral bending loads. The longerons have a rectangular cross section and consist of unidirectional glass rovings pultruded under tension through a bath of epoxy resin. The fuselage skins carry the shear loads due to vertical shear, lateral shear and torsion, and are stiffened by eleven internal glass fibre frames. Concentrated loads are input at some of these frames. Frame 1 reacts the engine frame loads, frames 3 and 4 are the main wing attachments and frames 10 and 11 provide the tailplane attachments. In order to prevent the relatively thin

S695/006/99 © IMechE 2000

glass laminate skins (typically 1 mm) from becoming unstable under shear loading, glasscloth top hat stiffeners and ring frames stiffen the fuselage. To stiffen the fuselage skins further against shear instability, the cloths are layed up at ±45° thus providing optimum shear strength and stiffness. Another reason for preventing instability in the airframe skins is that shear buckles can cause delamination in the glass laminates.

The wing also carries bending, shear and torsion loads and has two spars. The main front spar is continuous running from the port wing tip to the starboard wing tip. The rear spar is also continuous but runs from port rib 6 to starboard rib 6. The pilots effectively sit on the wing centre section. The wing main spar carries the wing bending loads with the spar caps carrying the wing bending moment as differential tension and compression endloads. The spar caps are manufactured from pultruded glass rovings in the same way as the fuselage longerons. The integral vertical shear web between the spar caps carries the flexural shear. The shear web consists of a birch ply and Rohacell foam core sandwiched by thin glass fibre skins layed up at ±45°. The birch ply is positioned at load input points such as the wing to fuselage attachments. The wing rear spar reacts the main landing gear loads. The rear spar shear web also consists of a birch ply and Rohacell core sandwiched by glass skins layed up at ±45°.

The wing upper and lower skins react the inplane shear loads due to torsion. The wing skins are stiffened with corrugations and ribs which are layed up at ±45° to optimise shear strength and stiffness. The corrugations run in a chordwise direction in order to stiffen the wing skins against shear instability only, and so as not to attract compressive loads. The wing is designed so that the main spar caps carry all the spanwise compressive loads generated by wing bending.

The eight ribs in the wing are positioned at the points where loads are input from the flaps, ailerons and main landing gear.

The tailplane is of similar construction to the wing. The tailplane skins are also stabilised to prevent shear instability, with corrugations running chordwise.

The aircraft is wet layed up in an open mould with glass cloth and Shell Epikote/Epikure 162/113 epoxy resin and hardener. Epoxy resins distort less under cure than do polyester resins and for this reason are used exclusively for aviation products. The wing skins and corrugations are consolidated using 5 Kg weights and the fuselage skins are consolidated by rolling. The smaller tailplane skins and corrugations are consolidated using vacuum bags.

The fuselage is manufactured in two halves with the skins being layed up at ±45° into each mould. The skins are then allowed to cure at room temperature. The frames are manufactured separately and are cured at room temperature before being bonded into one half of the fuselage with Redux epoxy adhesive. After room temperature cure of the Redux, the other half of the fuselage skin is bonded in place. The wing and tailplane skins are layed up in two open moulds at ±45° orientation. The corrugations are layed up in their moulds and whilst wet are layed onto the wet skins. Weights are placed on top of the corrugation moulds in order to provide consolidation during cure at room temperature. At this stage the pre-made ribs and spars are bonded in place with Redux. After room temperature cure the upper skin is bonded in place.

The fuselage, wing, tailplane and control surfaces are finally post cured in a hot box in order for the resin to achieve its full strength. The post curing cycle is 8 hours at 42°C and 16 hours at 78°C ±2°C. This results in a glass transition temperature of around 92°C.

Many current aircraft and glider fuselage and wing skins consist of foam cores sandwiched by thin glass or carbon skins, which removes the need for ring frames and stiffeners. Current foams are designed for high temperature, impact, and fire resistance. The Firefly was designed in the early nineteen eighties and the reason why the conventional stiffening approach was preferred at that time was that foam sandwich structures presented problems under impact loads. Some glass sandwich structures manufactured with earlier foams were impact tested, so as to represent the dropping of tools, with resulting friability problems. At the point of impact, damage occurred in the foam which then expanded as the laminate was cyclically loaded.

2.1 T67 cowlings

For the higher powered T67M260 Firefly variant with the 260 HP engine, it was necessary to reduce the weight of certain components. One of the components chosen for weight saving was the cowling, (Fig 2). The cowling is a non structural component which covers the engine and consists of upper and lower cowlings, which are bolted together. The whole assembly is cantilevered off the firewall at Frame 1. The cowling needs to be inherently stiff as it is positioned directly behind the propeller and if not adequately stiffened is prone to excessive deformation and cracking. As there is very little space between the engine components, baffles, sump, air intake box and the cowling skins then stiffeners have to be shallow. The lower cowling has inherent stiffness because of its tight radii and curvatures around the air intake box. The upper cowling, however has large flat areas that require extra stiffening.

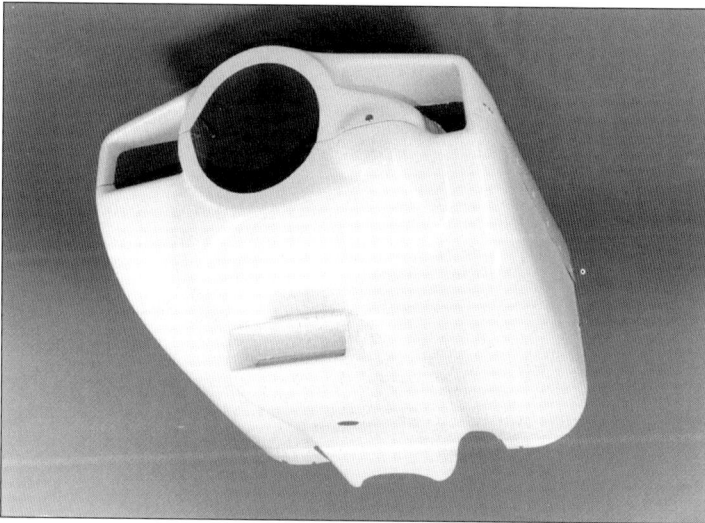

Figure 2 T67 Firefly cowlings

The original T67 cowlings were manufactured with wet layup glasscloth and Kevlar. Skin thicknesses of 3mm meant that internal stiffeners were not required.

S695/006/99 © IMechE 2000

In order to save weight and improve the inherent stiffness of the upper and lower cowlings, a hybrid glass/carbon/polyester prepreg, Vicotex 914/G973 was chosen. Vicotex914/G973 is stiffer than glass cloth. The cowling layup is 5 layers of Vicotex, with 3 layers at 0°/90° sandwiching 2 layers at ±45°. This gives an overall wall thickness of 1 mm. Shallow transverse stiffeners are needed to stiffen the flat upper and vertical surfaces of the cowling. This is accomplished with two 20 mm wide x 6.35 mm thick Nomex honeycomb cores, covered by the inner layer of the layup.

A 15 lb (40%) weight saving is achieved by this design when compared with the original wet layup cowling.

Positioned behind the propeller, the cowling is in a primary zone for lightning strike. A single glass cloth outer surface layer over the hybrid cloth (51.2% carbon, 34.8% E glass fibre and 14% polyester fibre), results in an 'equivalent glass structure' with respect to lightning strike. The only electrical bonding required was on the fasteners joining the two cowling halves.

3   JETSTREAM 41 BAGGAGE POD

The Jetstream 41 baggage pod (Fig 3) forms part of a ventral structure attached below the fuselage. The whole composite assembly consists of a forward fairing, a hydraulics bay, a battery support structure and the baggage pod itself. The baggage pod allows carriage of up to 350 lb of baggage.

Figure 3 Jetstream 41 baggage pod

The baggage pod consists of Fibrelam bulkheads which attach to the fuselage and a sandwich skin which is manufactured with a layer of glass and Kevlar on each side of a Nomex core (Fig 4). The Fibrelam is supplied as flat sheets and consists of a phenolic coated aramid honeycomb core sandwiched between unidirectional glass fibre reinforced epoxy resin skins.

The diagram shows a cross-section layup:

92125 Glass cloth 0°/90°
Kevlar 0°/90°
Nomex HRH 10/OX - 3/16 3.0
Kevlar 0°/90°
Kevlar ±45°
92125 Glass cloth 0°/90°

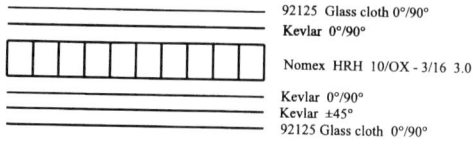

The Moisture Absorption of Jetstream 41 Baggage Pod Test Pieces At 85% Relative Humidity And 65°C

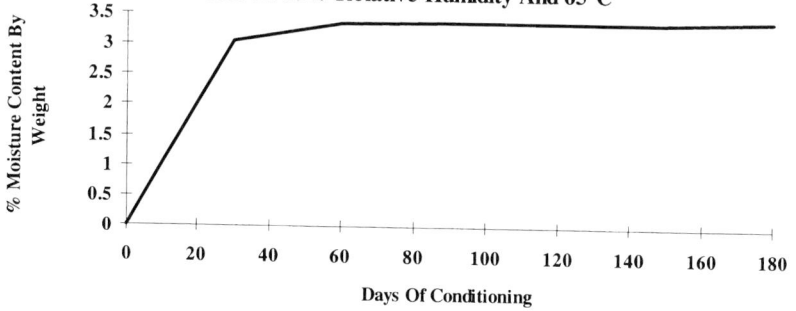

Y-axis: % Moisture Content By Weight (0, 0.5, 1, 1.5, 2, 2.5, 3, 3.5)

X-axis: Days Of Conditioning (0, 20, 40, 60, 80, 100, 120, 140, 160, 180)

Flexural Strength Of Baggage Pod Specimens At Various States Of Moisture Content And Temperature

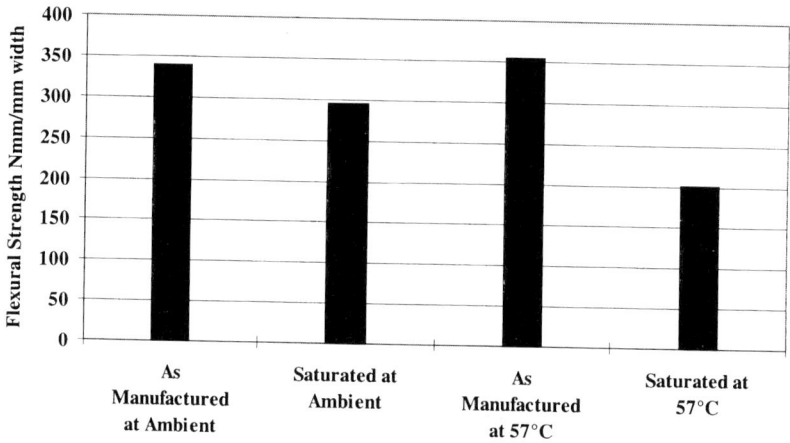

Y-axis: Flexural Strength Nmm/mm width (0, 50, 100, 150, 200, 250, 300, 350, 400)

X-axis categories:
- As Manufactured at Ambient
- Saturated at Ambient
- As Manufactured at 57°C
- Saturated at 57°C

Figure 4 Test results for flexural tests on baggage pod coupon specimens

The baggage pod skins are wet laid up in a mould with CTM epoxy resin. The skins are consolidated onto the core with vacuum bags and the assembly is post cured in a hot box with a post cure cycle of 12 hours at 78°C.

In order to stress check the baggage pod, a finite element model of the baggage pod was constructed. The shell of the pod and Fibrelam bulkheads were divided into 5 inch (127mm) wide strips. The honeycomb construction of the pod shell and bulkheads was then modelled as beam elements and shear panels. The beam properties represented the bending strength of the Fibrelam and the glass skin Nomex glass skin sandwich of the pod shell. The Nomex and Fibrelam cores were assumed to carry the through thickness flexural shear and the skins the bending moment. The shear panels between the beams represented the shear capability of the pod and bulkhead skins.

The baggage pod has been statically tested at ambient 'as manufactured', with Barely Visible Impact Damage (BVID) induced and delaminations built in (reference 1). The ultimate loads were factored up by 'hot/wet' factors to cover the pod for world-wide use.

The Kevlar, although excellent for impact and abrasion resistance is a hygroscopic aramid, the fibres of which absorb moisture, expanding in the process. The baggage pod was designed to operate world-wide, and Jetstream 41s are currently operating in the hot humid Far East climes. As part of the validation of the pod structure, it was necessary to validate the hot and wet factors used on the static test by long term tests on coupon specimens. To perform the validation, flexural coupon specimens were cut from a flat hydraulic access door which has the same construction as the pod. The specimens were saturated in a humidity cabinet. The worst case scenario is of a saturated baggage pod, at the certification temperature, and under critical load. Flexural testing was carried out on 'as manufactured' specimens at the certification temperature of 57°C. Test results (reference Fig 4) indicate that the saturated specimens at 57°C have a flexural strength 59% that of the 'as manufactured' strength at ambient. The test results indicated that under critical load and temperature, a saturated pod was of adequate strength with acceptable reserve.

4   VICKERS NAVAL GUN TURRET

The current naval gun turret consists of a curved turret manufactured with chop strand matt skins on each side of a balsa wood core, and a steel floor with steel support beams.

The redesigned composite turret has faceted sides with the floor and underfloor beams redesigned in composite. The overall size of the turret is 4.3 metres long, 3.2 metres wide and 3 metres high.

The latest design load specifications require structures mounted on the forecastle of ships to be able to resist peak wave slap pressures of 140 KPa for 15ms and 70 MPa for 350 ms. This loading represented a large increase compared to the loads under which the original design was stressed.

The turret skins consist of multiaxial stitched glass cloths 8mm thick sandwiching a 50mm thick closed cell foam core. The turret was manufactured by resin infusing the glass foam sandwich structure with vinyl ester resin in an open, inverted female mould. After manufacture the resin was cured at 70°C for 20 hours in a hot box.

As the turret structure consists of flat facets, pressure loads due to wave slap induce bending and well as membrane loads in the panels. To determine the strength of the panels under bending, a number of representative beams were manufactured (reference 3). These beams were 1.8 metres long and 100mm wide. The beams were flexurally tested on a three point bend test rig. The failure in all the beams occurred as a consequence of the interlaminar shear failure between the foam and the skins. No failure occurred in the glass skins. The beams failed at an average load of 7 KN, with a resulting maximum skin stress of 80 MPa. Although the skins did not fail at this stress level, the foam did and therefore the skin stress of 80 MPa was used as an indication of foam failure. The critical bending stress was taken as the mean of the test results divided by two in order to take into account the hot and wet degradation of the skins over time. The critical limiting stress in the turret skins was therefore taken to be 40 MPa. Following the flexural test, the interlaminar shear strength of the beams was determined. The interlaminar shear test consisted of loading the beams on a three point bend test rig with a span to thickness ratio of 5 to 1. The interlaminar shear strength of the beams was around 3.5 MPa

A finite element analysis was then performed on the turret and floor. The finite element model (reference Fig 5) represented the turret skins as laminate elements containing the outer and inner skins and foam core. The outer skins and inner skins were modelled as orthotropic plate elements and the foam cores as isotropic elements. The stiffness properties input into the model (Young's modulii, shear modulii and Poisson's ratio) were derived by testing coupon specimens. Manufacturer's stiffness values were input for the foam.

The finite element model indicated that the maximum bending stresses in the turret skins were less than the critical limiting 40 MPa which would indicate delamination of the foam core.

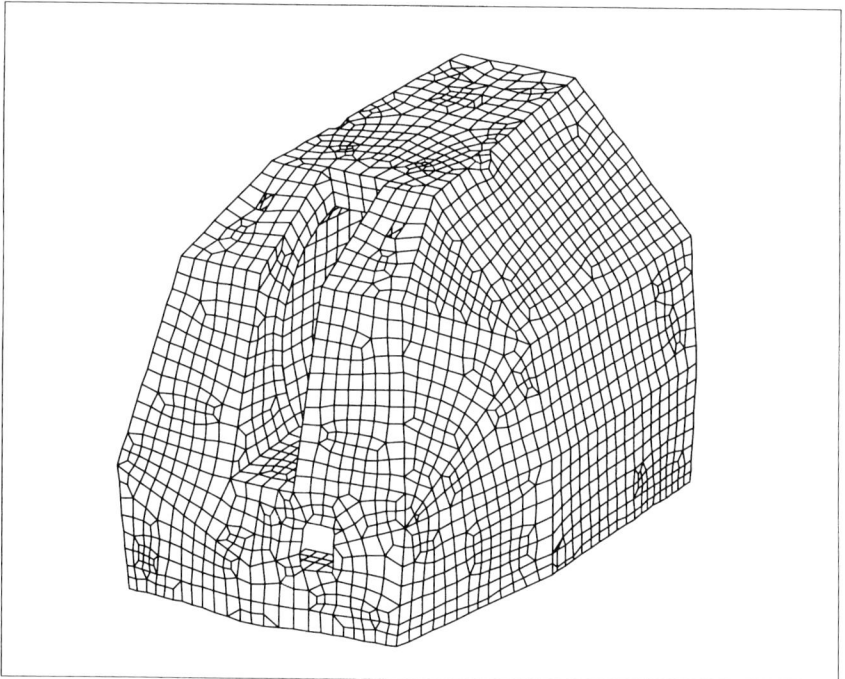

Figure 5 Naval gun turret finite element model

S695/006/99 © IMechE 2000

5   DOWTY R408 SPINNER CONE

The Dowty R408 spinner cone is attached to the spinner backplate which in turn is attached to the propeller hub.

The spinner cone stiffness is derived from it's shape and must maintain that shape with negligible deformation during operation, especially during overspeed. The spinner was originally manufactured from aluminium and as part of a weight saving programme, the spinner was redesigned in prepreg glasscloth.   This resulted in a weight saving of 9.4 lbs which represents 40% of the weight of the original aluminium spinner.

The spinner consists of two parts, a cone body and an internal diaphragm which provides extra stiffness. The circular diaphragm, bonded inside and at the mid length of the spinner, consists of two layers of cloth each side of a 3 mm thick honeycomb core. A forward extension of the hub passes through the centre of the diaphragm, thus providing a support.

In addition to maintaining its shape during operation, the spinner cone has to resist birdstrike without jeopardising the operation of the aircraft. The spinner is also in a primary zone for lightning strike.

Due to the above design parameters, the spinner is manufactured using MTM48 glass prepreg (MTM48-GF0100-42%Vf-40%RC) with an alumesh outer layer to provide continuity for lightning strike.

MTM48 is a tough flexible epoxy resin system. The tensile modulus of MTM48 at 0°/90° is 21,000 MPa which compares with 33,200 MPa for 920G glass prepreg.

The spinner cone consists of five layers of MTM48 layed up at ±45° orientation, resulting in a cone thickness of 1.3 mm. Aft of the diaphragm, the spinner cone thickens to 15 layers of MTM48 in order to react the bearing loads at the attachment bolts. The prepreg is cured in an autoclave at 125°C and 3 bar pressure.

The spinner cone was mounted on a production carbon prepreg backplate and subjected to a bird strike test (reference 4). A JAR 4 lb chicken was fired at the spinner cone, striking the cone at a speed of 255 knots (294 mph). The damage (reference Fig 6) consisted of a large dent with two small cracks, the detachment of the diaphragm which was pushed back, and the loss of 6 out of 12 of the attachment fasteners. The bird penetrated the carbon backplate causing it to shatter into a number of pieces as can be seen in the photograph (Fig 6). This test is a good illustration of the relative impact strengths of glass fibre and carbon fibre. Following the bird strike test, the damaged spinner cone was subjected to 850 revs per minute for 20 minutes, which it survived with no further damage.

Figure 6 Dowty R408 spinner cone after bird strike test

6   SUBMARINE COMPOSITE RUDDER

Naval ship and submarine rudders are normally free flood structures and consist of thin steel plates welded to an underlying steel framework. These types of structure are notoriously prone to accelerated corrosion because the internal volume is inaccessible and stays wet for a long time. Typically submarine rudders require some remedial action every two years. On one submarine in refit, the rudder had to be completely rebuilt.

Effective solutions to the corrosion problem have been found for ships. Sandwich panels of steel and light weight syntactic foam are known to provide cost and weight effective solutions for rudders. For a submarine rudder, foams capable of withstanding deep dive pressures exist and could be used with a steel skin. They would be weight effective submerged but rather heavy in air, a particular problem for a tall upper rudder.

The composite rudder was developed not only to solve the rudder corrosion problem but also to spearhead a wider application of composites on submarines. Hydroplanes, ducts casings and fins are all free flood structures prone to corrosion. Fibre reinforced plastic skins on underlying steel frames had previously been used for top casings and fins. The rudder was chosen as the "all FRP" proving ground because it is a large safety critical structure, with risk mitigated by fairly easy reversion to a steel rudder if it failed.

6.1 Rudder specification

a)   A structure predominantly FRP, ie free choice of fibre matrix or layup.

b)   All external dimensions as current rudder.

c)   Service and environmental load capability as for current rudder.

d)   Hydrodynamic smoothness maintained under load.

e)   Capability for retrofit to an existing steel shaft on an in service submarine.

f)   Easy through life inspection and monitoring of prototype until proven.

g)   Prototype build should cost no more than current cost of a steel rudder.

h)   Negligible through life cost and maintenance requirements.

i)   Weight no greater than current rudder in air or submerged.

6.2 Design parameters

The rudder is designed to be resin infused. Resin infusion can be used for single skin and core constructions. The resulting composite material properties directly compare to properties that have only been achieved with prepregs in highly controlled, expensive autoclave processes. Dry cloths are layed into an open mould and resin is drawn through the laminate under vacuum. Resin infusion is a repeatable process which improves the fibre/resin content of the laminate and as a consequence the strength and stiffness. Void content is reduced and manufacture is less labour intensive than hand wet lay up. Once the resin content has reached equilibrium, (70% fibre content by weight is easily achievable) the process then stops.

Because of the marine environment, vinyl ester resin was chosen. Derakane vinyl ester was chosen because of its suitability for resin infusion. For cost reasons the skin and the ribs are manufactured using glasscloth.

The rudder has to be capable of reacting large side forces and transferring the resulting shear, torsion and bending to the submarine structure. Internal ribs transfer shear across to the submarine pintle, the glasscloth skins reacting the torsional loads by Bredt Batho shear. A finite element model of the rudder, showing the internal structural components, is shown in Fig 7.

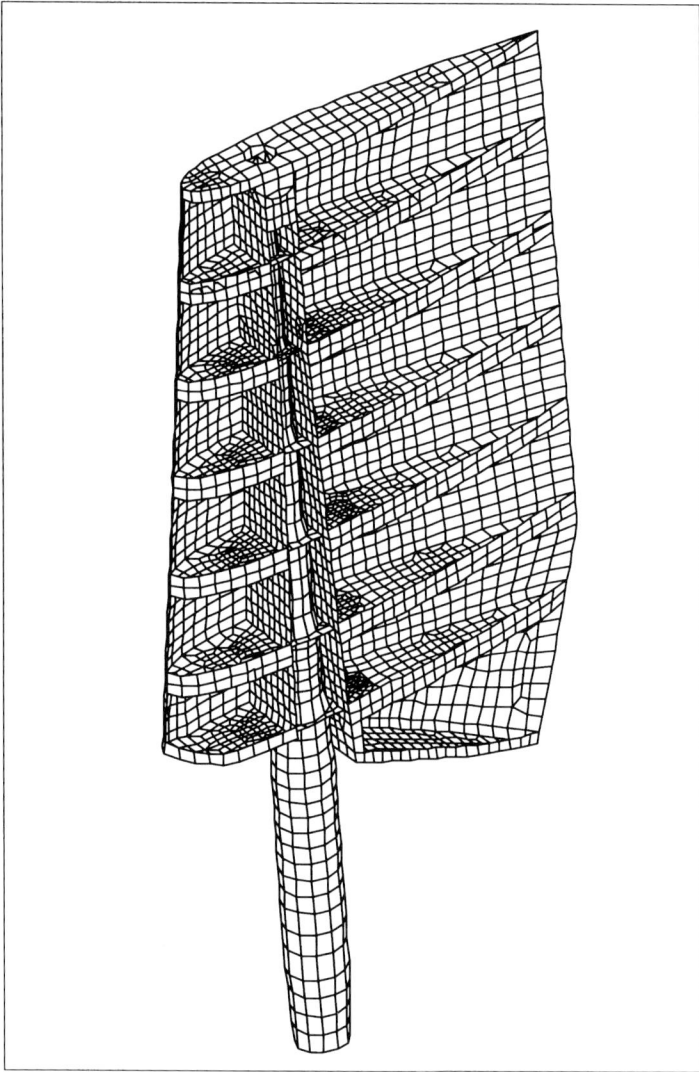

Figure 7 Finite element model of composite rudder

The rudder skins consist of two relatively thin laminates separated by a greater thickness of anechoic medium. The laminates consist of alternative layers of biaxial 0°/90° and ±45° stitched fabrics. As the inner skin is directly attached to the ribs, it is assumed to carry all the torsional shear in the rudder. Likewise the inner skins react the spanwise and transverse bending loads as endload. The outer skin is effectively isolated by the anechoic medium and is therefore assumed to only react bending loads due to the panels between the ribs reacting pressure.

The ribs consist of back to back 'C' channel sections forming an 'I' cross section. The rib webs have holes in them to allow the flooding of the rudder. The ribs have a pitch that gives the skins an acceptable span for reacting out of plane pressure loads. As the rudder is subject to internal pressure loads in service this means that the skin and rib flange joints will be put into peel. Peel loads have been avoided in the rest of the rudder design, shear being the preferred load transfer mechanism through the joints. Where peel is unavoidable, the joint tensile stresses are designed to be low. For this reason the rib has an 'I' cross section giving a maximised skin to rib flange bond area.

7   HOVERCRAFT HULL DESIGN

The Slingsby SAH 2200 hovercraft (Fig 8) in its standard configuration, is outfitted for 2 crew and 21 passengers, and is capable of speeds up to 40 knots. In the military role this configuration can be altered to suit operator requirements. The hovercraft is constructed in composite materials because this gives significant benefits in corrosion resistance, fatigue life and low maintenance requirements. The undersides are reinforced with Kevlar. The craft are powered by Deutz turbo charged air cooled engines. The hovercraft have been operated by the Royal Marines, UK Ministry of defence, Finnish and Swedish coast guard.

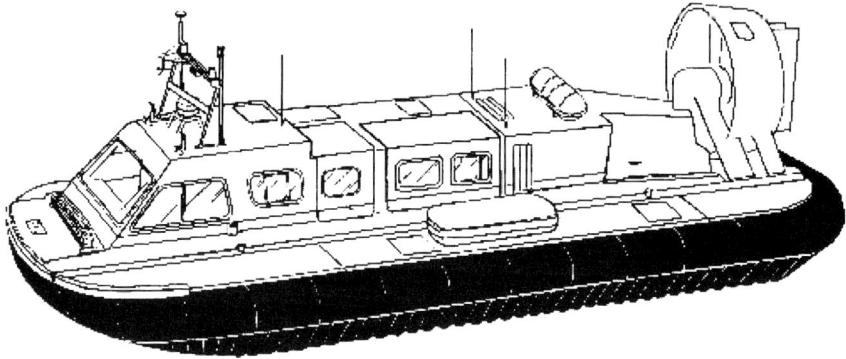

Figure 8 SAH 2200 hovercraft

The Slingsby SAH2200 hovercraft is manufactured using hand wet layup techniques. The hull floors, side walls and roofs consist of thin glass skins on each side of a foam core. Three craft are currently being supplied to Finland and the design requirement was that the floor of the craft should react ice pinnacle loads. In its operational environment, the craft may come to rest on ice pinnacles. Research into ice columns has indicated that pressures of up to 6.5 MPa can be generated on the bottom of the craft before the ice column starts to break up (reference

5). The standard SAH2200 cabin floor is not reinforced to carry the high 6.5 MPa ice pressure loads. The cabin floor is a glass/foam/glass sandwich. In order to reinforce the floor, a keel beam and transverse beams were introduced between the upper and lower skins. The introduction of the keel and transverse beams had the effect of reducing the panel sizes and hence increased the flexural strength of the floor. The lower skin was further thickened to improve the laminar shear strength of the impacted surface. The keel and transverse beam arrangement is shown in Fig 9.

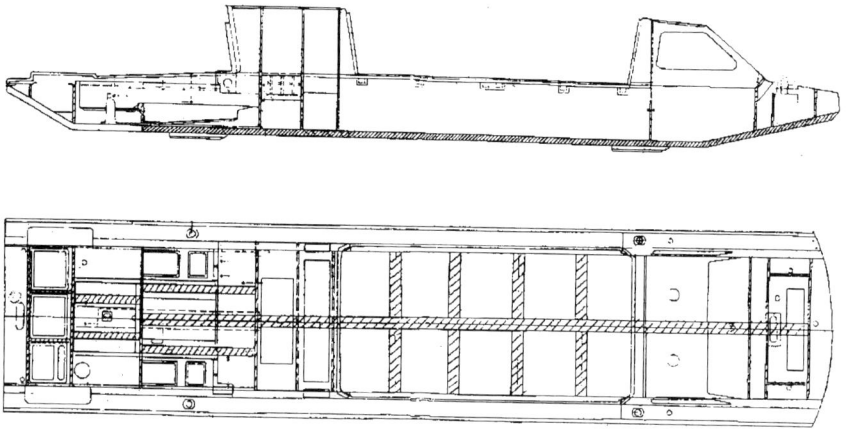

Figure 9 Keel beam and transverse beam arrangement in the hovercraft floor

The required thickness of the lower skins was derived empirically (reference 6). Sandwich panels with the same layup and dimensions as the actual floor panels, were tested with various lower skin layups. A Denison test machine applied the ice loading and the ice pinnacle was represented by a cylindrical steel penetrator with a diameter of 100 mm. To allow the test panel to flex, the edges of the panel were supported on wooden blocks. This ensured that the mode of failure was representative of a combination of interlaminar shear and flexure. The typical failure mechanism was a combination of the shear failure in the foam as shown in Fig 10 and a through thickness shear failure of the lower skin around the periphery of the penetrator.

S695/006/99 © IMechE 2000

Figure 10 Ice pinnacle load test with floor panel in denison

8 REFERENCES

1   Slingsby Test Report   Jetstream 41 Baggage Pod Static Strength Test SAL STR058 (1992).

2   Slingsby Test Report   Baggage Pod Pickup Point Fatigue Test   SAL R&DTR 348 (1992).

3   Slingsby Test Report   'The flexural and ILS strengths and the resin/glass content determinations for sandwich laminates manufactured with EQX2336 quadriaxial glasscloth and Derakane resin'. SAL R&DTR 723 (October 1998).

4   Slingsby Test Report 'Bird Impact Testing Of The R408 Composite Propeller Spinner'. AE06-950-001 (December 1997).

5   **Tapio Nyman** Resistance And Manoeuvring Tests In Ice Of The Ice Breaker Sampo.

     VTT Technical Research Centre Of Finland ESPOO (1995).

6   Slingsby Test Report 'Further tests to determine the ice loads and pressures required to cause failure of the floor lower skin of Finnish hovercraft numbers 28, 29 and 30'. SAL R&DTR720 (September 1998).

7   **D.J. Cresswell, J House, N. Foster** Composite Rudder Steers Into The Future!

     International Conference, Advances In Marine Structures III, Dunfermline Dunfermline Conference Proceedings Volume 1 Papers 1 – 21. (20 – 23 May 1997).